ROUTLEDGE LIBRARY EDITIONS: EVOLUTION

Volume 6

NATURE AND HISTORY

NATURE AND HISTORY
The Evolutionary Approach for Social Scientists

IGNAZIO MASULLI

Routledge
Taylor & Francis Group

LONDON AND NEW YORK

First published in 1990 by Gordon & Breach Science Publishers

This edition first published in 2020
by Routledge
2 Park Square, Milton Park, Abingdon, Oxon OX14 4RN

and by Routledge
52 Vanderbilt Avenue, New York, NY 10017

Routledge is an imprint of the Taylor & Francis Group, an informa business

© 1990 OPA (Amsterdam) B.V.

British Library Cataloguing in Publication Data
A catalogue record for this book is available from the British Library

ISBN: 978-0-367-27938-7 (Set)
ISBN: 978-0-429-31628-9 (Set) (ebk)
ISBN: 978-0-367-26036-1 (Volume 6) (hbk)
ISBN: 978-0-367-26079-8 (Volume 6) (pbk)
ISBN: 978-0-429-29136-4 (Volume 6) (ebk)

Publisher's Note
The publisher has gone to great lengths to ensure the quality of this reprint but points out that some imperfections in the original copies may be apparent.

Disclaimer
The publisher has made every effort to trace copyright holders and would welcome correspondence from those they have been unable to trace.

NATURE AND HISTORY

The Evolutionary Approach for Social Scientists

by

Ignazio Masulli
University of Bologna
Italy

Gordon and Breach Science Publishers
New York Philadelphia London Paris Montreux Tokyo Melbourne

Copyright © 1990 by OPA (Amsterdam) B.V. All rights reserved. Published under license by Gordon and Breach Science Publishers S.A.

Gordon and Breach Science Publishers

Post Office Box 786
Cooper Station
New York, New York 10276
United States of America

5301 Tacony Street, Drawer 330
Philadelphia, Pennsylvania 19137
United States of America

Post Office Box 197
London WC2E 9PX
United Kingdom

58, rue Lhomond
75005 Paris
France

Post Office Box 161
1820 Montreux 2
Switzerland

3–14–9, Okubo
Shinjuku-ku, Tokyo 169
Japan

Private Bag 8
Camberwell, Victoria 3124
Australia

Library of Congress Cataloging-in-Publication Data

Masulli, Ignazio.
    Nature and history : the evolutionary approach for social scientists / by Ignazio Masulli.
        p.   cm. — (The World futures general evolution studies : v. 1)
    Includes bibliographical references.
    ISBN 2–88124–376–2 (Switzerland)
        1. Social history—Methodology.   2. Social sciences—Methodology.
    3. Science—Methodology.   4. System theory.   I. Title.   II. Series.
HN28.M37   1990
300'.72—dc20
                            90–3082
                               CIP

CONTENTS

CONTENTS

ACKNOWLEDGMENTS

This book originated from methodological reflections on the socio-historical sciences as compared with the natural sciences. This necessitated a long interdisciplinary research which led me to epistemological conclusions of a more general nature.

The chance to discuss the results of my research with Ervin Laszlo has been very important. I owe him a debt of gratitude for much more than his invaluable advice and encouragement. Meeting him has been one of the greatest privileges of my intellectual life.

I should also like to thank Timothy Keates for the patience and dedication with which he faced the numerous difficulties in translating my text.

PART ONE:

The paradigm of form

CHAPTER 1

Synthesis of Biological and Psychological Processes

The last few decades have witnessed some astonishing advances and achievements in science. This has brought about a situation where we can no longer view the evolution of nature and of man, his behavior, his form of organization, or his very thought, in the more traditional, more stock terms of our culture. Yet again science has forged ahead far beyond the conceptions we cling to, of ourselves, our lives; and this advance has come about not merely in the common way of such things, it pushes beyond many of the categories still predominant in the organization of scientific knowledge. The very grammar of this knowledge is changing before our eyes, demanding that we adopt new paradigms.

Several of the findings of contemporary science have contributed to new applications and substantiations of a basic question that has never been absent from the background and has often been touched upon or openly mooted. Today it can once again be raised in all its theoretical implications.

Can evolution, progressing by stages and degrees, from the simplest to the most complex forms, be subjected to a "historical reading"? So numerous and of such moment are the contributions of recent developments in the sciences, that even as regards the two points of evolution where the qualitative leap seemed at its most striking—that of the origins of life and that of human thought—we may assert with Jacob that today "physics dissolves in the study of the cell as biology dissolves in the study of man".[1]

The findings and developments of molecular biology, genetics, embryology, population genetics, the biology of evolution and palaeontology, on the one hand, and the achievements of neurophysiology, the biology of knowledge, genetic psychology and the theory of cognition on the

3

other, have led to an extensive reappraisal of the phenomena and conceptions regarding man's natural and cultural evolution. An important part has also been played in these achievements by the so-called "transverse disciplines": cybernetics, systems theory, information theory.

The emergence of new perspectives and new scientific paths in seemingly more distant research fields, such as non-equilibrium thermodynamics, non-linear chemistry, and the dynamic systems theories in the area of theoretical mathematics, not to speak of the social sciences and cosmology—all have lent further scope and diversification to the overall picture.

On the basis of this scientific pluralism the problem has been posed as to how to establish an epistemology of complexity that shall be capable of accounting for "those articulations that are destroyed by the separations between disciplines, between cognitive categories and types of knowledge"[2].

The study of complex systems, independently of whatever specific field of investigation they belong to, now enables us to deal in more general terms with the problems of evolution[3]. But we must be ready to abandon several of our more traditional perspectives and points of reference. Among these the relationship between biological evolution and man's cultural evolution is foremost in posing the greatest problems of conceptual revision.

These are no longer questions arising out of the Darwinian perspective, however viable that perspective may still substantially be. We can no longer rest content merely to trace the history of human evolution, of the determining of superior traits in the evolution of the human species that characterize it in some special way and mark it off from some other species or other outcomes, different yet contiguous, in the course of evolution. The problems are quite new ones, engendered by the latest findings of biological, cognitive and systems sciences; and even as they repropound the problem of the relation between the biological and the cultural evolution of man, so they shift this problem to a different qualitative plane.

To sum up, we stand at the crossroads of epistemological problems and the findings in various research fields—in biology, psychology, sociology and the sciences of cognition[4].

The relation between two levels of man's evolution represents a junction at which several new research hypotheses converge and from which new paths branch out towards different results.

In the meanwhile, the terms "biological" and "cultural" are themselves in process of thorough-going redefinition. On the one hand, the biological elements reveal themselves to be of ever greater consequence and function, so much so as to break the bounds within which they were traditionally confined. On the other hand, the cultural elements, when analyzed as to certain of their essential and recurrent characteristics, disclose such deep substrata of acquisition and elaboration as to require thorough reappraisal of the patterns and significances of the demarcations that have traditionally circumscribed that "other" realm, the realm of culture.

Far-reaching and radical these redefinitions may be; but it is in the study of these new connexions that we ultimately come face to face with the problem of the relationship between the two levels of evolution, the biological and the cultural.

How far should we consider biological evolution as the "basis" for explanatory models of man's behavior and cognitive functions? In what way do the genetic "dictate", the heredity "historically" amassed by the species and the refinements continually performed at the first level, open out onto the second level? And what is their relation to the evolution of man's thought and action on the second, the cultural level?

The genetic dictate and the heredity amassed by the species clearly cannot be considered as a finished datum in itself, something on the "basis" of which is grafted and unfolded the current, eminently cultural history of man's latest achievements in knowledge and organizations; in which case, how should we view the "vertical" axis of the cultural-biological relationship? There is, too, another so-called "horizontal" dimension, which also belongs with the traditional way of conceiving the relation between the two levels: it sees cultural evolution as history, as having followed a biological evolution now concluded—how should we reappraise this in the light of the latest findings of science?

As regards both dimensions of the relationship, it can be seen that there is plenty of scope for presupposing a solution of continuity from the biological level to the cultural level. But each time this solution is once more posited, and however it is advanced, it results in a non-explanation, a sort of *a posteriori* annullment of that relationship; ultimately, a recoil from examining and reappraising all those linkages between man's cultural and biological dimensions towards which so many of the latest developments in science have pointed. And this leads to the positing of an eminently theoretical problem, that of reformulating the terms of the relationship.

Piaget rightly remarked that the "reconstruction" performed by human knowledge, starting from the bursting of instinct, is actually so thorough-going that it has kept many theorists of knowledge aloof from the need to explain it "by going back to the obviously necessary frameworks of the living organization"[5].

But cybernetics, advances in neurophysiology and new approaches to the problem of knowledge, interwoven with the expansion and elaboration of topics in the field of evolutionary biology, were to underline this need once again. And already by the mid-sixties a whole series of findings which the study of evolution had amassed had to be taken into account[6].

On the basis of these premises, in their general implications, and to the extent that the hereditary tools of knowledge were more directly referred to, Piaget was able to go ahead with his biological interpretation of the higher forms of knowledge.

Cognitive regulation began by using the tools generally employed by organic adaptation: heredity and phenotypic accommodation. But with the final emergence of instinct,—which leads to a dissociation of its two components, internal organization and phenotypic accommodation—we get a complementary reconstruction in opposite directions, leading to the dual formation of logical-mathematical structures and experimental knowledge[7].

Thus Piaget can assert that, as far as its content is concerned, thought starts from scratch, but it is functionally prepared, not only by sensory-motor and nervous coordinations but also, and more fundamentally, by everything that the nervous functions have, in turn, inherited from the organic functions as a whole. "It must, indeed, be clearly understood that the general organizational conditions that we proposed. . . . as a possible basis for logico-mathematical structurations are not, chronologically speaking, initial ones, but generalized and at work all the time"[8].

Thus the mechanisms which they (general conditions of organization) determine can very well serve as objects for the reflective abstractions that characterize thoughts. In other words, logical-mathematical structures extend the general organizing function common to all living systems, both because this function is present in action and in the nervous system, and because the reflective abstractions cannot be said to have an absolute beginning but, rather, can be retraced to the "convergent reconstructions with overtakings"[9].

As for experimental knowledge (a higher form of learning), it differs from, but is closely linked with, logical-mathematical knowledge. This is very important in as much as "to say that physical knowledge is an assimilation of the real world into logico-mathematical structures amounts, in fact, to affirming. . . . that the organization belonging to a subject or to any living creature is a condition of exchanges with environment and cognitive exchanges, just as much as it is a condition of material and energy exchanges. In this respect, conceptual and operational "forms" appear yet again as the extension of organic "forms"[10].

Formulations like these were later to be resumed or reworked in research and in theoretical elaboration. To this we shall return later on in the book. But as regards the elements considered hitherto, it must already be apparent that the configuration of the relation between the biological and the cultural levels of evolution may be seen, quite wrongly and unsatisfactorily, as one of subordination, whereby the one level constitutes a "basis" for the other, with an irremovable demarcation between them.

In actual fact, the acquisitions, the elaborations, the achievements at biological level interact and link up with cultural findings at every point and passage of human evolution. Far from being a precisely delimited datum, no longer subject to change, or merely underlying, the biological level not only influences or acts upon the cultural level but actually opens onto the entire unravelling that may be possible at cultural level[11].

Just as the acquisitions and elaborations of the cultural level comprehend the biological level, i.e. include a biological content; and this content comprehends and unfolds in the evolution that pertains to it. In other words, cultural evolution is also, simultaneously, biological; as biological evolution is also, simultaneously, cultural.

The second, as it were "horizontal" dimension of the relationship has rather to do with the continuity from that which precedes to that which follows in man's biological and cultural evolution. In this case, too, traditional logic, purely linear and consequential, delimiting what precedes in terms of what follows, may conceive it as *minus habens* and may easily establish an improper hierarchy in the horizontal and temporal dimensions of evolution as well.

Once again we find ourselves face to face with a "mysterious" solution of continuity: the relationship is not really explained and the continuity is interrupted only because to the second level is assigned what is denied to the first. Or so it would be, were it not for the fact that, once again, the latest findings and tendencies of research suggest that the continuity between the two levels of human evolution has a much greater significance.

The ever more obvious relevance in human thought and action of all that is implied in the heritage of our species—an evolving heritage, not merely acquired and transmitted but also constantly enriched at each stage of evolution—calls for the complete revision of the merely consequential relation between the first and second levels. To put it another way, we can no longer conceive of a relationship between biological and cultural evolution such that, "at a certain point", the "historical" result of what took place on the first level determined the conditions necessary for the second level to "begin".

Here, too, within this second dimension of the relationship, that which takes place on the first level can be seen as opening out onto the second. As the importance of the role played by biological agents in the unfolding of cognitive and behavioral functions becomes increasingly evident, we cannot underestimate the ability of these agents to predispose to thoughts and actions; and this is a predisposition that is not single, once and for all, but rather continually elaborated and refashioned— in short, active at every point in cultural evolution. It is wholly connected with that evolution, leaving no room for distinctions between before and after, between preceding and following.

We must thus fully acknowledge the importance of the self-construction, biological and cultural, of a thought, an action, a choice. And it still remains to draw all the theoretical consequences of a statement of this kind, such as the present state of our knowledge makes possible[12].

Every distinction between before and after collapses in the relation between the two levels, and the predisposing capacity cannot be confined to the biological level nor be seen as concluded there; and thus, that which is commonly understood as belonging to the second, cultural level of human evolution—i.e. the capacity to acquire new elements—is itself already active at the first level.

Still more important, this capacity and function can be said not only to belong also to the biological level of evolution, but actually to stand in direct relation to the same capacity and function of the second level; or as I should prefer to put it, to be *open* to the second level.

Needless to say, neither cultural nor biological evolution could exist, were it not for this capacity-tendency to acquire or to learn something new; it is this very capacity-tendency that demonstrates more than any other thing how the evolution of the first level is also, and simultaneously, that of the second, and vice versa.

To be sure, if we would avoid a reductionist conception of biological phenomena and of the relation between the two levels of human evolution (reductive in any sense, including that of ascribing a super-determining character to what is biologically given), we must have sufficient understanding of how biological agents constantly incline towards learning: that is to say, they incline towards recognition of what is new and towards relating themselves to it, by virtue of the reasons behind their own evolution. Moreover, their tendency is to appropriate the new unto themselves and to change with it, in a process that fits perfectly into the same logic as governs the cognitive processes at the second level.

Here, too, as regards the capacity for learning, every distinction between a before and an after in the relation between the two levels disappears.

But if biological evolution, both in its capacity for predisposition and its capacity for learning, proves open to cultural evolution, and that which occurs on the first level also partakes of the second level and vice versa, then biological evolution cannot be viewed as a long sequence preceding cultural evolution: we cannot see the point of arrival of the former as constituting the point of departure of the latter; or, to put it another way, that the results determined at the first level have, *at a certain point*, made possible the activities of the second level.

In the relation between the biological and cultural levels of evolution, then, we cannot establish a delimiting distinction—according to traditional logic—between what is below and what is above, or between what precedes and what follows as a consequence. And, as we have seen, in both the vertical and horizontal dimensions of the relationship the first level partakes of the second and vice versa. Granting all this, we must conclude that each of the two levels comprehends the other.

And this path leads us to acknowledge a bio-psychic synthesis in evolution.

In the history of scientific thought there is nothing new in the positing of a synthesis of this kind. Certain of its premises can be found in the problems of knowledge arising out of developments in microphysics. Indeed, in that same field of scientific enquiry the relation between subject and object of investigation is radically altered: to the effect that the former cannot properly be distinguished from the latter, cannot be considered external to it. This intrinsic, determining character of the subject with respect to the findings of scientific observation was pointed out by Wolfgang Pauli, who went so far as to indicate a genuinely psychological-subjective correspondence in scientific explanation. Pauli actually regarded this correspondence as an *initial condition* if scientific explanation was to be rationally satisfied[13]. From this point of view, he pondered the "innate mental tendencies of man", according to which man judges certain rational connexions of explicative character between intimate and external facts to be satisfactory[14].

And it is highly significant that Pauli explicitly drew on the developments of biology and psychology and the findings that were already emerging in those years: this led him to assert that, thanks to new and parallel discoveries in those two fields, a substantially different view of evolution was affirming itself. This new view demanded that an area of interrelation between unconscious psyche and physical processes be taken into account[15]. And along these lines Pauli conclusively asserted the "one-ness of the physical and the psychic spheres of reality".

It is clear that, subsequent to Pauli's thought, findings in biology and psychology, and in fields where these two traditional disciplines encroach on one another, have shown in more precise, more cogent terms that a distinction between the two spheres of reality to which two substantially separate levels of evolution might be assigned can no longer be maintained.

Nevertheless, Pauli's statement remains important for the very reason that each time we investigate the relation between the two levels of human evolution, biological and cultural, and try to single out a connexion beyond mere relation as understood in terms of traditional logic, the conception of two distinct spheres of reality re-emerges, even in the new formulations. The penalty for not overcoming this conception is that each nexus loses some of its significance.

On the other hand, it is obvious that a different conception in which each level of evolution partakes of the other and includes it—leading, as we have seen, to the assertion of a bio-psychic synthesis of human evolution—cannot, in the last analysis, avoid implying the "oneness of the two spheres, physical and psychic, of reality". In other words, it cannot help alluding to a more general synthesis in which evolution itself may find all the terms for its own explanation.

As a result of the annullment of all distinctions and delimitation between "spheres", we can then fully understand how the evolutionary process is open at both levels; i.e. how the evolutionary process on the biological level is open to the cultural level and vice versa.

Indeed, in the bio-psychic synthesis itself we may posit the open character of the entire evolutionary process. (This overrides every distinction or delimitation, which are ultimately self-contradictory and limiting).

Now, to assert that the evolutionary process is wholly open means that the predisposition inherent in it does not constitute a dictate, already defined and distinct, or enjoying a status of priority with respect to learning (the acquisition of new elements). Rather should this predisposition be understood, more precisely, as predisposition towards learning and learning as a renewal of predisposition. In other words, predisposition aims at what is to be learned which is, in turn, learned according to a predisposition immediately renewed in the very act of learning.

Between the twin terms a permanent, symmetrical tension is set up and this underlies each outcome-passage in evolutionary change.

If what evolves is a *form*, evolving according to form, it must be stated that this form is nothing other than a tension, in so far as it is a tending towards a form and, at the same time, a form forming that same tendency.

Form is thus the active synthesis that complies with the bio-psychic synthesis in a process open to both levels of evolution. And the tension that characterizes *form* is of psychoid nature[16]. Which is to say, we have here a tension both psychic and physical; not a purely psychological, but rather a psychic tension of physical breadth and content (and vice versa). It is the tension of a synthesis that corresponds to the psycho-physical one-ness of reality and expresses that one-ness in the concrete unfolding of evolution.

Which means that this tension—inclining towards a form that is already form—urges form to partake of all the elements towards which it tends and to comprehend them through the psychoid tension that represents its real mode of unfolding.

NOTES TO CHAPTER 1

1. Jacob (1970).
2. Morin (1984).
3. Laszlo (1986).
4. Various studies, significant in different ways, bear on this problem, which in the last analysis implies a redefinition of the very characteristics of knowledge itself: such are, for instance, the studies by McCulloch, Piaget, von Foerster, Bateson, Maturana, Varela, Atlan. On the inescapability of the tangle of problems arising in the cognitive and biological sciences, see Rossi (1981), pp. 403–428.
5. Piaget (1967, English trans. 1971), p. 367.
6. The elaborations of Dobzhansky, Wallace and Lerner on the concept of genetic pool, those of Darlington and Lewontin on genetic recombination, the formulations of Dobzhansky and Waddington of the problem of the relation between phenotypical adjustments and genetic adaptation—all these were elements that could be combined in a reappraisal of the problem raised by classical theories. In Piaget's view, these difficulties were especially serious inasmuch as they regarded the need for reciprocity between the structures of the genome and its morphogenetic activities. Nonetheless, it began to appear that the problems might be solved, thanks to procedures taken from cybernetics—procedures able to transcend one-way linear causality; thanks also to the importance assumed by regulation mechanisms in the new lines of research. The latter could already be found in the studies of Waddington and White.
7. Piaget (1967, English trans. 1971), pp. 364–5.
8. *Ibid.*, p. 333.
9. *Ibid., p. 333.*

10. *Ibid.*, pp. 338–9.
11. After stressing the "merely partial" rigidity of the genetic programme, Jacob goes on to say: "The procedure can be described as a tending towards elasticity in carrying out the programme in a sense that enables the organism to acquire autonomy, to build up relations with the environment and thus extend its range of action and activities. This progressive opening of the programme manifests itself especially in the tools available to the organism for collecting information from outside, for processing that information and for reacting accordingly" (Jacob, 1974).
12. A new understanding of the biological nature of the structures of thought was facilitated by the new directions taken by both neurology and genetics. These had already begun to appear in the mid-1960s and were to be further developed. I shall have more to say about these developments later on. In the field of neurology, starting with Jackson and Sherrington at the latest, attention shifted to the constructive processes characterized by the integration of inferior structures in those at higher level, passing through successive stages of formation. This passage featured a reorganization and recombination of elements already present at the preceding levels, though in a less differentiated way. Parallel with this, there was a growing interest in the problems of regulation (re-combination, re-equilibrium and organization). This led to a better understanding of the phenomena of recombination, in which connexion variation was no longer to be referred merely to fortuitous mutations but also to re-equilibrium of the whole system. So that one must bear in mind the dual nature of the process of transmission: alongside the transmission of genetic information, an agent in the morphogenesis of the generation coming after, self-conservation must also be considered— that is to say, conservation by self-regulation of the organization itself and of the transmission of its mode of functioning in the course of its multiple activities. In this perspective, the functioning that continues during the transmissions does not transmit itself in the true sense of the term, but can be considered a basic transmission inasmuch as it *remains constantly active*. So that, from the point of view of the biology of knowledge we can say that learning and predisposition find a preliminary and necessary condition in "the organizing function with its absolute continuity". The organizing function should thus be understood as the "necessary condition for any transmission and vice versa" (Piaget, 1967, English trans. 1971), p. 322.
13. Pauli (1961), pp. 113–17.
14. Jung-Pauli (1952) p. 163, *passim*. On the relation between subject and object of observation I shall have more to say later, and shall deal with the substantial modifications of that relationship in physics and biology, as also from the standpoint of cognition. As regards Pauli's formulation of this relationship and its implications, in the terms in which he approaches it, an interesting comparison can be made with the standpoint of the biology of knowledge. The agreement between schemes of knowledge—which in the last analysis reflect the forms of organization of life—and their contents is actuated not from the outside, nor by some pre-established harmony, but rather within the organism itself. This requires that we go back to the source of the general coordination of the operation performed on the environment. But although it is held that the source of this coordination (and the structures of thought to which it leads) lies in the more general functioning of the living organization, a more general agreement between this coordination or structures and the external environment is allowed (Piaget, 1967, English trans. pp. 344–5). Thus we reach a conclusion substantially similar to Pauli's. And the analogy will be all the more evident when we compare Piaget's and Pauli's approaches to the problem of understanding mathemat-

ics. Pauli speaks of "original mathematical intuition" and asserts: "Even when we confront the greatest achievements of the human spirit, like mathematics, we must not forget the continuity of life, which reconnects the emergence of concepts with the phenomena of adaptation of all living organisms" (Pauli, 1961, p. 122).

15. This concept can easily be derived from a group of remarks by Pauli on biological evolution (*Ibid.*, pp. 123–25). For all sorts of reasons these remarks seem highly relevant.

16. As we know, the concept of "psychoid" was employed by Jung (1954) to take into account the non-psychic. My way of understanding it, however, refers to the interpretation put forward by Pauli. What Pauli found promising in Jungian psychology was its evolution towards the "recognition of the non-psychic in relation to the problem of unity of mind and body". Referring to the, from this point of view, most significant formulations of Jung, Pauli concluded: "Personally, I see here the first sign of recognition of an ordering principle, that is neutral with respect to the 'psychic-physical distinction' "; and he added that this principle was to be understood abstractly, "not intuitively in and for itself" (Pauli, 1961, pp. 118–22). His conclusion was ultimately, and openly, consistent with the problem of observation, from which his thinking on this subject started. Thus we see how this problem had travelled from microphysics onto a more general plane, but without losing sight of its scientific basis.

CHAPTER 2

The Nature-Knowledge Nexus and the Synthesis of Form

In dealing with human evolution it may be somewhat easier to understand and demonstrate the bio-psychic synthesis that characterizes it; as we have seen, this may lead us once again to locate the act of cognition within the totality of the process of organization of the living being— even at the level of the most differentiated and specialized evolution of cognitive regulation compared with organic regulation, as can be found in human evolution. This re-insertion must, *a fortiori,* be performed at other levels of biological evolution and in terms of a more general bio-psychic synthesis.

The basic hypotheses of a biology of knowledge concern, in effect, the general terms of the regulations in the organization of the living being. At the various levels of biological evolution the cognitive regulations "are an extension of organic regulations", of which they represent an 'outcome'; so that we may say that cognitive functions "constitute a differentiated organ for regulating exchanges with the external world"[1]. And this involves going beyond any traditional reference to the subject of knowledge. Indeed, if the real is understandable only in terms of the organization of reality, such organization cannot be referred back to the human subject alone: that would carry the risk of anthropocentrism "with minimal gain", an anthropocentrism that has saddled philosophy with a problem of the "subject" that can hardly be solved in traditional terms. But if "the very nature of life is constantly to overtake itself, and if we seek the explanation of rational organization within the living organization, *including its overtakings,* we are attempting to interpret knowledge in terms of its own construction, which is no longer an absurd method since knowledge is *essentially construction*"[2]. Now it is just this constructivist hypothesis, informing Piaget's genetic epistemology[3], that on the one hand has contributed to restating the need for a relationship between epistemology, biology and other sciences, as I mentioned in the opening pages; on the other hand, it has been taken up and elaborated in the course of the subsequent development of the sciences of evolution and cognition.

From our point of view, it is important to focus on the change that has come about—in the wake of the aforesaid developments—in the conception of the relation between organism and environment. As we have seen, the theoretical problem for organization began to be clear enough in the period marked by the *Biology of Knowledge,* and this necessity was destined to assume considerable importance in the study of organic and cognitive systems. And indeed it is on the very basis of this problem that the relationship between cognitive regulations and organic regulations has been established, in terms of the realization of "the internal program of the organization in general"[4]; and the organizational "closing" of the system (which is open to its environment) enables its environment to be progressively extended, both in the biological and the cognitive senses. So that the essential functions of the cognitive mechanisms consist, on the one hand, in progressive closure of the open system of the organism—thanks to an indefinite extension of the environment—and, on the other hand, in the equilibrium mechanisms of the system. If living organization is essentially self-regulation, the development of the cognitive functions consists in the construction of specialized regulating organs that serve to govern exchanges with the exterior and, ultimately, to enable the functioning of living organization as such[5].

The formulations inherent in the *Biology of Knowledge* already represent a considerable shift in the way of looking at the terms of the relationship between organism and environment: this no longer appears as a mere relationship of adaptation characterized by a pure and simple response of the organism to environmental requirements. Rather, what emerges is a much more complex circle of exchanges—characterized by interaction and feedback—such as to admit of a selection of environmental stimuli on the part of the organism and a capacity to respond to these stimuli and endow them with significance; a capacity to respond aimed at maintaining the organism's own organization, even in continual adaptation. This, as we have seen, means a precise relation between organic and cognitive regulation, referable to the whole living organization.

Thanks to subsequent progress in systems theory, a perspective of this kind was not only confirmed but further enlarged. These developments were accompanied by a whole series of achievements and pointed the way along various paths of research in evolutionary biology and neurophysiology.

Generally speaking, there is a move away from a conception of control of the system on the basis of external necessities towards another conception in which reference to its internal structure and maintenance of its autonomy takes pride of place, including all the possible interactions between system-and-system and system-and-environment. Indeed, it is the system's very ability to enter into interaction without loss of autonomy that leads us to define the *domain of interactions of the system as its cognitive domain*[6].

This new shift in the conspectus—highly relevant to the problem we are dealing with here—was supported with new results and new lines of research in the sectors most nearly concerned in the redefining of relations between system and environment and of the resulting view of adaptation. Important in this connexion was the work done in evolutionary biology, such as that of Gould and Lewontin[7], and the establishment of a new line of research in the neurological sciences, where the concept of system autonomy was resumed and elaborated[8]. Adaptation was thus analyzed in terms of compatibility between structure of environment and structure of system; and it was this compatibility that led to real structural coupling between system and environment: reciprocal disturbances cause continual changes in state and the structural couplings are inherent in those changes[9].

The result is a notably different configuration—not only of the nervous system but of knowledge itself. By now this latter is no longer conceived as a process of optimization on the basis of correspondence with an external norm. Attention shifts from the exterior to the interior: it becomes of central importance to consider the mode in which the structure of a system produces compensation for the interactions. Most interesting of all, from the point of view of knowledge, is the change of structure. The cognitive process involves a change in structure in the continuous maintenance of integrity of the system[10]. Knowledge is no longer regarded from the point of view of representation; rather, it is understood essentially as construction[11].

At which point, the abstract problem of the "origin" of knowledge no longer holds; there is no point in adducing some *sui generis* reference, "mental" or otherwise[12]. From the constructivist standpoint there is no gap to be bridged.

In the new terms of a consistently constructivist epistemology it may be stated that "living systems are cognitive systems" and that "living, as a process, is a process of cognition"[13]. And this holds good for all organisms, with or without a nervous system. The nervous system enlarges the domain of interaction of the organism, so that its internal states can be modified even by "pure relations" as well as by physical events. Moreover, by thus enlarging the organism's domain of interaction, the nervous system subjects the acting and interacting that occur in the domain of "pure relations" to a process of evolution[14].

Thus we are in possession of all the necessary elements to conclude that knowledge organizes itself in a way corresponding to the organization of natural processes of evolution; the advances it involves are demanded by that selfsame evolutionary progress. And in this way the fully organic character of the nature-knowledge nexus is reconfirmed.

This organicity means, if anything, that knowledge is re-inscribed in natural evolution.

Thus, as Piaget's constructivist hypothesis is resumed and elaborated, we find that his important caveat as to the risk of a philosophical anthropocentrism deriving from the traditional subjective-transcendent reference regarding the source of knowledge, becomes much more than a mere caveat.

Indeed, one of the most significant epistemological consequences of the change in perspective—from representation to construction—such as has come about in the biological, neurological and cognitive sciences, is that man cannot step outside his own cognitive domain (thus distinguishing himself in subjective terms, or in terms of subject alone)—that domain specified by his identity and integrity as a living system. In effect, he cannot choose where that domain shall begin, nor its modalities[15].

But the full re-insertion of knowledge in natural evolution has nothing less than the significance of a nature that, by evolving, knows itself.

By its properties, *form* shows itself to be the active synthesis presiding over the concrete unfolding of the nature-knowledge nexus at each degree and passage of the evolutionary process.

So that we can say that at each degree of the process of evolution a form of that process obtains—on the one hand, recognizable, on the other corresponding to that degree, or form (for this reason the cognizability is assumed in the process itself *qua* form); moreover, each form knows itself and its own process from within, in as much as it is not the result of that process but rather its property or form.

The form of a process is the qualitative and functional property already wholly inscribed in that form (and without which the process cannot take place). The *form* is an understanding which is also a performing, a performing which is also an understanding. In it, what is understood is performed in as much as it is understood, and vice versa: this is active synthesis which, at each passage in the evolutionary process, is an understanding and performing together.

In *form* the twin terms "nature" and "knowledge" do not merely stand in a simple determinate relationship to each other: rather, the organicity of their interrelationship constitutes an intimate nexus.

In form, as active synthesis of the evolution process, nature and knowledge proceed as one along the evolutionary scale.

That it is the properties of form that strongly characterize the nature-knowledge nexus is demonstrated by the theoretical implications of the very concept of form.

Indeed, form can in no way be reduced to object (formed form); to put it another way, it cannot be conceived as the result of a performance and an understanding whose function and significance are distinct from the result. Nor can form be reduced to subject, with a distinction of function and significance repropounded on this other plane in similar fashion.

Form is neither subject nor object of any process (natural and cognitive) that can be located outside of it (distinguishable by itself), since it contains (within itself) all the properties and natural and cognitive qualities of the evolutionary process of which it constitutes the active synthesis.

If form expresses the nature-knowledge nexus at the various degrees of evolution, then a systematic character can be detected in that nexus from the simplest forms to the most complex; the latter occurring in human evolution and expressing the highest degree of knowledge. But it is clear that such knowledge emerges only by virtue of the systematicity of the nature-knowledge nexus, and without that systematicity no knowledge would actually be possible: this because it would be quite abstract and "external"—which is to say, literally *placed outside*.

Form is thus not the object of knowledge: it is itself *logos,* the language of nature, or nature-knowledge.

So that we can speak of a knowledge wholly comprised in natural evolution and belonging to a nature that, in evolving, knows itself.

Whence we can derive some theoretical consequences of no small moment. In the first place, only the intimate connexion between nature and knowledge, as established by the synthesis of form, guarantees the knowability of the real (a knowability in itself otherwise abstract). Otherwise (outside that nexus) one would always be compelled to search, at philosophical level, for a special perspective, of phenomenological or ontological type, in order to close a gap or compensate for an alienness that would, in any case, persist.

In sum, the concept of *form* supplies a theoretical foundation for a scientific knowledge finally freed from the problems of the subjectivity and objectivity of knowledge that have traditionally beset it; and we can then proceed towards the conception of a total knowledge—total in so far as relinked to the general process of evolution. At the same time, the conception of a nature that, in evolving, knows itself, gives rise to a second and equally relevant consideration: if the highest degree of this evolution involves the knowability of man, this cannot be isolated or reduced to subjectivity.

The overcoming of every "humanistic" abstraction, which has located the source of all knowledge within man as a mere subject, means a restoration of wholeness also to that same human knowledge (relinking it to the general process of evolution from which and with which it proceeds).

Beginning once more from total man as an expression of the nature-knowledge nexus, means healing all those wounds that derive from the conception of partial man—all the ways in which that partiality has been conceived—and that have troubled science[16]. The most obvious manifestation of this is in the separation of "the two cultures": the one dealing with the sciences of nature, the other with the sciences of man. This is the most striking example of the irremediable rupture brought about by an entirely arbitrary distinction-separation set up as a starting-point. It is reflected in all the logical and epistemological partialities that stem from an abstract conception of what is knowable.

Redirecting human thought towards the nature-knowledge nexus in the process of evolution, comprising it in the synthesis of that nexus as constituted in *form,* means liberating it from a conception of knowledge as something detached, abstracted from the total reality of man; it means restoring man to the whole complexity of his being and his evolution.

For several elements of the sterility, alienation and inadequacy of our culture derive from that mutilated and abstract conception of man. Hence the need to rethink man's very psychology, his behavior in history, his culture—with which latter, after all, he constructs his own future. The danger in such a state of cultural alienation lies in the fact that man is constructing his future in a manner at variance with his own self, with nature, with his deepest tendencies.

So that it becomes imperative to correct this abstraction, this "humanistic" alienation at a time when man can intervene in his own genetic inheritance and in the overall equilibria of nature. This redress is, moveover, necessitated by the progress of the sciences, since they offer man the instruments for the exercise of a power and a will that require to be thought out anew and freshly understood in a realer, wider perspective; and because they bear witness to the systematicity of the nature-knowledge nexus in an evolution of which man is the outcome. The development of science thus suggests a new mode of thinking and gives man the devices by which to enact it.

And since we cannot pass beyond a particular level of culture and knowledge except as the evolution of that level requires, so we find ourselves face to face with a new "consciousness", a new elaboration of human thought, all of whose possibilities and necessities now appear.

But who can doubt that this qualitative leap in the evolution-knowledge of man already involves a new revolution in scientific knowledge—one of the same scope as the "humanist" revolution which gave birth to the new science of the XVII century?

The start of that scientific revolution saw a fresh conception of man, of the reality of his nature and his cognitive ability. With the *De humani corporis fabrica* of Andreas Vesalius, man was conceived as a secularized object for investigation: man submitted himself for study as a natural objective datum. And thus there was born within him that subject-object division in knowledge, that man-versus nature separation that mirrored the basis of the humanistic prejudice. Man subjected himself to "anatomical" study of himself in order to discover the possibilities of the "human machine". According to the logic of that view, the next step could only be the Galilean telescope: that is to say, the search for prosthetic instruments necessary for increasing man's power and knowledge. Hence the birth of the new science, functional, operative, geometrical, as deriving from man's mode of conceiving the study of man. But hence, also, the defect in that nature-knowledge separation, and the resulting view of the dominion of man (as subject of knowledge) over nature (as object of knowledge).

This conception has been hard to combat and difficult to overcome, even today when all the necessary conditions for going beyond it exist; and this is reflected in the risk incurred of imposing a traditional interpretation on the new findings in science, with all the foreseeable consequences of hedging them about with reductivist restrictions.

In the afterglow of these remarkable achievements in molecular biology, genetics and neurophysiology, we may be tempted to reformulate a new "anatomy": this is what tends to happen when such findings are interpreted in a reductionist way by referring to single components and functions, thus losing sight of the intimate and highly significant interconnexions they themselves suggest[17].

But however scrupulous we may be in not reiterating the traditional conceptions and logical schemes in the reading and conceptual systematization of scientific results (even the most highly innovative ones), there is always the almost unavoidable danger that they will loom up again where experimental research is not accompanied by theoretical elaboration adequate to the scope of the problems and questions raised by that research; and this means that the need for theorization has not been sufficiently met[18].

Thus, even in the sectors I have mentioned, it can and does happen that with the vast number of different lines of research and the rapid expansion of these, a growing specialization and sectorialization become necessary; but they carry with them the penalty of narrowness. The greatest danger is that under these conditions an explanation of the phenomena investigated may confine itself to the terms of reference pertaining to the particular narrow field in question. And here the risk of reductionism is inherent.

On the contrary, we see that the most cogent and significant results of contemporary biology—and, as we may say, its maturity—have been attained in the fundamental stages by reference and relation to other disciplines. This has opened the way to fresh theory.

In this connexion, the scientific itinerary of Watson and Crick in the discovery of the structure of DNA is well known. Their prevailing interest, according to them, was in theory, arising out of hints from quantum physics, and especially from the questions posed by Schroedinger as to genetic functioning. What they did, in effect, was to apply the informationist, structurist and biochemical approaches to the problems of genetics.

Following the climax that molecular biology reached with the discovery of the structure of DNA, the next phase, which also produced some astonishing findings, was marked by an urgent need for theorization.

In the present situation of biological studies and theories of evolution we face fundamental questions that, on the one hand, represent a synthesis of the current state of knowledge and, on the other, express the need for theory arising from the very results achieved—theory without which significant research can hardly proceed. Here we come to the heart of the problems.

Not by chance, one of these has to do with embryology. The genoma has a one-dimensional linear organization given by the genes arranged in linear series in the chromosomes. How is this organization translated into the three-dimensional one of the organism? In this connexion, Jacob states that the sole logic that biologists have been able to master is one-dimensional: "If molecular biology was able to develop rapidly, it was largely because information in biology happens to be determined by linear sequences of building blocks." Whereas what takes place in the development of the embryo is "the production of two-dimensional cell layers that fold in a precise way to produce the three-dimensional tissues and organs that give the organism its shape and properties"[19].

The difficulty consists in passing from the anatomy of the molecule to the way in which the instruction is given for building an organ and an organism.

And, linked with the first, a second problem emerges: what are the mechanisms by which the cells "destined" to form a certain organ recognize each other as the embryo proceeds to form? With problems of this kind we reach the real threshold of an understanding of life.

Here is a good demonstration of the need to perform an authentic mental "leap", such as that necessary to comprehend the living system in terms of its *organization*.

The difficulties derive from the fact that in the past attention was predominantly focused on aspects of information (transmission, conservation, expression)[20]. But these difficulties are also of a conceptual kind, where "we have no proper ways of thinking about the organization at all"[21].

And, in effect, the problem of organization raises a real demand for theory, coming at a certain point in the progress of research and requiring to be interpreted if this research is to take a new direction.

It concerns, at one and the same time, a new conception of the living system and of the evolutionary process: and it is significant that it should arise at the point where the two planes cross.

In any case, the findings of research have, of themselves, given us an ever more complex, more problematic vision of the living system and the relationship between the problems of regulation and development and those of evolution.

The stratified configuration of the living system, the complexity of the relations between part and whole, the problems of regulation, as dealt with in most recent studies[22], offer the most significant points of reference; and the same may be said of the more rigorous systematizations in neurophysiology. These results also invite us, once more, to tackle the important problems already posed in the past[23].

Moreover, it becomes clearer and clearer that embryology has a contribution to make to the study of evolution[24].

In addition, the identification of the gaps that require to be closed in our knowledge of regulatory circuits and mechanisms of embryonic development, together with a growing awareness of their importance, have played no small part in the reformulation of the problems[25].

What finally emerges is a new dimension of the evolution of organization and regulation[26]; this converges with a new systemic and stratified conception of the organism in the reconsideration of the various mechanisms and levels in the evolutionary process.

But the prospect of broadening and diversifying the factors and mechanisms of evolution arises from the encounter between the needs arising out of research into evolution and the changes in epistemology in the study of complex systems.

The formulations of the study of these systems invite contemporary biology to reappraise the relations between the parts and the whole, and between the system and the environment. And this clearly favors a positive interpretation of the need for organization, as emerges from the results of research. It also enables us to resume the study of historical problems which were formerly relegated to a lowly status by the old conspectus, and more precisely by its limited and reductivist definitions; according to this constricted view, pride of place was to be awarded to an evolutionary biology dealing with parts and genes rather than organisms, thus undervaluing the integrated systems of development[27]. Instead, the new perspective restores full significance to the overall organization of the morphological, physiological and behavioral systems to which the various parts belong, since the variability of these latter depends on that organization no less than on the demands of the environment[28].

It is thus possible to conceive of an effective circularity in the system-environment relationship, in which the first term is not simply a passive one[29].

Pursuing the lines and suggestions of this research we get a glimpse of a wider horizon for convergence, for a broader and deeper approach to the living system, its interactions, its domains, its organic and cognitive regulation.

At bottom, as we have seen, the new perspective implies and supports a totally different formulation of the system-environment relationship. The connotation of living system as cognitive system, which can be constated on this basis, concerns the totality of the process of evolution and the nature-knowledge nexus by which it is characterized.

Thus if research into problems of evolution finds itself compelled to employ new terms of reference, the findings of that research cannot but support, at theoretical level, the presence of a self-organization inherent in the living system: a self-organization that admits and demands the systematicity of the nature-knowledge nexus in the evolutionary process. Form—as active synthesis in which that nexus is expressed—then appears as a new paradigm able to represent the general terms of the evolutionary process.

In the meantime, empirical research proceeds at various levels—including the strictly molecular-biological—along its own lines, dealing with specific questions (such as the study of the external membranes of cells, the identification of the molecules whose function it is to recognize cells, the functioning of hormones in recognizing and binding molecules produced by other organs and in the specific functions they exert on other cells).

The results so far achieved, or likely to be achieved, are so important as to have, in their turn, new implications for theory. For this very reason, however, it is essential to avoid reductivist interpretations. And we may fall into this very trap if, failing the mental "leap" required by our interpretation of the demand for theory of organization, we retreat to "atomistic" constructions of those results. Nor can there be any doubt but that this kind of interpretation leads to dogmatic banalization of the various popularized socio-biological gospels of recent years[30].

Thus the most traditional conception of the man-nature relationship is turned upside down. But the same old defect as before emerges (though with a mirror effect), and it is instructive to observe how it can assume the connotations of a species of revived anthropomorphism. In these cases we ultimately find ourselves bogged down in the age-old antitheses—antitheses which can only be resolved with the reference to the nature-knowledge nexus.

It is the strong nexus of nature and knowledge—as expressed in the synthesis of form—that enables evolution to be conceived in new terms: that is to say, in terms of a nature that in evolving knows itself, from the most elementary level to the highest, i.e. that of the knowledge of man.

Thus a new conception of man and his relation with nature emerges, of no less importance than that which came into being in the period from the anatomists to Galileo and marked the birth of the "new science".

And we have every reason to believe that this is a real scientific revolution, in view of the vast importance and far-reaching implications in the astonishing findings of contemporary biology and its broad relations with other sciences (chemistry, physics, neurophysiology) which these findings have fostered and tend more and more to enhance—so as to breach at several points the wall that has hitherto divided the natural and the human sciences.

Indeed, the revolution involves the entire organization of the scientific system. Now, this scientific revolution, accompanied as it is by a new conception of the man-nature relationship, and having that new conception as its most striking manifestation, may represent a passing beyond that "humanistic" conception that has marked Western culture in modern times. But the process cannot consider itself as realized merely by affirming an opposite conception, i.e. the old one simply turned back to front and just as stale as before. In this way one might be tempted, for instance, towards a sort of "pan-biologism".

Instead, we find ourselves facing, if we will, a much more arduous, more complex vision: that of man as no longer the seat of an abstract rationality, set against nature, but rather capable of rethinking himself and nature *from within*—able to reconceive his own totality in fresh terms and with nothing left out of the picture.

NOTES TO CHAPTER 2

1. Piaget (1967, English trans. 1971), pp. 26 and 369.
2. *Ibid.*, p. 362.
3. On this connotation, which is fundamental in Piaget's research, see the later works, especially those of 1970 and 1975. On the relations with the other sciences see Piaget-Garcia (1983). The significance of Piaget's formulations in the panorama of contemporary epistemology has been amply expounded by Ceruti (1986), pp. 64 *et. seq.*
4. Piaget (1967, English trans. 1971), p. 364.
5. The analysis of the organization of organic and cognitive systems is resumed and further elaborated by Piaget in his books of 1974, 1975 and 1976.
6. In this connexion see the work of Varela-Goguen (1978), Varela (1979), Maturana-Varela (1980), Zeleny (1981), Maturana-Varela (1985).
7. Cf. Gould (1977a), (1980), (1983) and Gould-Lewontin (1979).
8. Maturan-Varela (1980).
9. *Ibid.*, p. xxi *passim.*
10. Varela (1984a), p. 219.
11. Ceruti (1985), p. 38.
12. Already, fifteen years ago, Young could remark that, in the current state of scientific knowledge, there was no longer any point in continuing to argue over an abstract "problem of mind": further advances in research would soon cause the problem to "vanish" (Young, 1971), p. 139.
13. Maturana-Varela (1980), p. 13.
14. *Ibid.*, p. xxi *passim.*
15. Varela, (1984b), p. 320.
16. Among the various critical approaches to this problem I quote, merely by way of example, the psychoanalytical approach of Brown (1959), the philosophical one of Whitehead (1954), and the social one of Marx (1844).
17. In this connexion see the critical remarks of Gould-Lewontin (1979).
18. Cf. Thom (1980), pp. 46–52 and 74–79.
19. Jacob (1983), p. 141.
20. Brenner (1987).
21. Young (1971), p. 74.

22. Ho-Saunders (1979).
23. Waddington (1975).
24. Gould (1985).
25. Jacob (1983), p. 141.
26. And according to Jacob this is the path we should take if we would discover evolution's real modes of proceeding; if it is true—as it surely is—that "what distinguishes a butterfly from a lion . . . is much less a difference in chemical constituents than in the organization and distribution of these constituents." *Ibid.*, p. 141.
27. Gould-Lewontin (1979), p. 597.
28. Bocchi-Ceruti (1984), p. 62.
29. Cf. Lewontin-Levins (1978) and Lewontin (1983).
30. On the dangers arising out of sociobiological interpretations, which are easily associated with "atomistic" dogma, see the repeated criticisms by Gould (1977a), Gould-Lewontin (1979), Lewontin (1978) and (1979). These studies should make clear the intrinsic limitations of a dogmatic faith in the possibility of "objectively" subdividing the organism into separate characters, thereby excluding its overall organization and the participation of the observer. But if we would overcome these limitations, we must renounce all exclusively adaptationist explanations of the evolutionary characters, as has been emphasized by Lewontin (1977).

CHAPTER 3

Nature and History: The Problem of a Unitary Conception of Reality

At this point it may be as well to take a long step backwards, in order to re-examine certain aspect of Engels's attempt to propound a new conception of the man-nature relationship as against the conception that held sway in the science of his time. Over and above the problem of "dialectic", central as this was for Engels (i.e. the dialectical terms of the new view he announced), was this approach to the findings and problems of science in his age. This is what will concern us here.

The reference to Engels is valid since the demand for theory as he saw it emerging from scientific research was intended to transfer itself to "the dimension within science"[1]. On the one hand this constitutes a methodological requisite of undeniable importance; on the other, Engels saw that the findings of science must be squarely faced up to. And this, in turn, led him to identify specific perspectives and formulations which are of particular relevance to my argument. It is well known that the basic hypothesis of Engelsian theory was that there is a substantial unity of thought and nature, such that the laws of the one are generated by the other and reflect its characteristics. Engels's purpose, therefore, was to foster the raising of new problems within the sciences, and in particular the problem of the universal connexion of everything that constituted reality. So that it was at precisely this point that there arose a myriad problems bearing on the state of sciences he proposed to deal with.

Meanwhile, his attempt could not but solicit an internal crisis within the sciences, concerning their theoretical foundations. And "the crisis could be set off only by challenging the autonomy of the different levels of reality that were presupposed, i.e. by calling for enquiry not only into the defined and autonomous areas of reality, but also into those moments of passage and unification that might serve to throw doubt on the definiteness and autonomy of those areas''[2]. Can we doubt that in its time, and the more as time went on, this was to prove the nub of a problem of the greatest significance? In connexion with a new historical and unitary conception of reality which aimed at destroying the static contiguity of the various levels of reality, there arose the problem of the continuity of the history of nature and the history of man. But this continuity, which had been suggested to Engels by the Darwinian theory of evolution (though with such additions and integrations as Engels felt necessary)[3], raised difficulties, both as regarded scientific findings and theory.

First and foremost, as Engels himself saw it, the validity of that perspective required that the progress of the sciences lead to the transcendence of the two basic solutions of continuity which were the most important in the moments of transition and unification envisaged by him: the transition from inorganic world to organic world, and from organic world to human thought.

But already Engels saw clearly that such progress must and could be afforded essentially by biology: which is to say, by a new domination on the part of the organic sciences, and by the new kind of relations they would be able to establish with the other sciences. And there can be no doubt but that the limitations in Engels's dealings with scientific developments derive from the historical inadequacies of these latter as much as from the terms of reference he chose to employ.

The state of the sciences, as referred to by Engels, actually presented a whole order of problems that could not but reflect themselves in the terms of his theoretical elaboration. The difficulties that Engels's approach encountered had to do with the definition of the modes of continuity. And it was, indeed, the indication of the modes of continuity that, more than any other problem, involved direct analysis of the findings and lines of scientific research in his time. This is the point where we can best gauge the distance that separates us from his onslaught on the question, and appreciate the novelty of the formulation of problems which the findings of research and their implications for theory have made possible.

Generally speaking, the interest with which Engels—from his particular viewpoint—regarded the progress of science in his time was determined by the fact that about half way through the XIX century "empirical natural science made such an advance and arrived at such brilliant results that not only did it become possible to overcome completely the mechanical one-sidedness of the eighteenth century, but also natural science itself, owing to the proof of the interconnections existing in nature itself between the various fields of investigation (mechanics, physics, chemistry, biology, etc.), was transformed from an empirical into a theoretical science and, by generalising the results achieved, into a system of the materialist knowledge of nature"[4].

And in the process science had achieved three results that he deemed of paramount importance and which enabled the historical continuity of natural evolution to be conceived.

The first was to have demonstrated the transformation of energy. This showed how "all the innumerable acting causes in nature" hitherto defined as "so-called forces" had now been proved to be "special forms, modes of existence of one and the same energy, i.e. motion".

The second result was due to the discovery "of the organic cell by Schwann and Schleiden, as being the unit out of which . . . all organisms, with the exception of the lowest, are formed and develop . . . The origin, growth and structure of organisms were deprived of their mysterious character".

But alongside these a third achievement, the most important, was necessary—namely, Darwin's theory of evolution: this in order that we might understand "the evolutionary series of organisms from a few simple forms to increasingly multifarious and complicated ones . . . extending right up to man." In this way, "the basis has also been provided for the prehistory of the human mind . . . " without which "the existence of the thinking human brain remains a miracle"[5].

To be sure, the ladder of this continuity was lacking one rung: the explanation of the "origin of life from inorganic nature"; but Engels had no doubt that further progress in physics and chemistry would provide the solution to the problem. So that he felt able to state that "today the whole of nature lies spread out before us, as a system of interconnexions and processes that, at least in all its main features, has been explained and understood"[6].

And it was precisely that "system of interconnexions and processes"—in other words, the ways in which the continuity could be conceived—that became the main focus of Engels's thought.

Engels identified the modes of continuity in the "forms of motion". Now, if we would speak of "general laws" of nature, uniformly applicable for "all bodies—from the nebula to man—", it must be concluded that "nothing remains as absolutely universally valid except motion."

To know the forms of motion, therefore, was the only way to know "*pro tanto*. . . . matter and motion *as such*"[7]. The very idea of matter thus appears to Engels as a nexus of the forms of motion. And "on the hypothesis that all the levels of matter are transmutable (their transformability thus suggesting a line of research) Engels's materialism takes its stand"[8].

But it was necessary to transcend a purely mechanical conception of motion. "Motion is not merely change of place: in fields higher than mechanics it is also change of quality"[9]. At which point the Engelsian notion of motion not only passes beyond mechanistic conceptions but, as it embraces each change of quality, so the category of form of motion can be referred to each level of reality and every possible transmutability. Indeed, "motion in the most general sense, conceived as the mode of existence, the inherent attribute, of matter, comprehends all changes and processes occurring in the universe, from mere change of place right up to thinking"[10].

The abstract model of matter thus derived does not, however, imply reduction to a single language, but rather the embracing of all the possible forms of motion and of the different languages they represent. Or rather, the various forms of motion are associated with each other and this association implies a unifying tension, that which is "historically possible"[11].

Substantially Engels, on the one hand, insists on emphasizing the specificity of the levels of reality in which the various forms of motion inhere and, on the other hand, stresses how they are of necessity connected with each other. And given the systematic character of these nexuses, it is possible to glimpse the nexuses that actually exist and those that, in the present state of science, appear only as points for future research or as problem situations[12].

In this way a theoretical perspective was outlined. It comprehended all the "continuity" implied by the Engelsian hypothesis. It was up to future science to demonstrate that continuity (on the basis of which the unitary conception of reality was advanced, in historical terms) by setting in motion those processes of linkage that had not as yet been contemplated; and this bore generally on everything that had to do with forms in the organic world.

If this was the perspective to which his theoretical model was committed, Engels was well aware of the limits of its applicability to the state of knowledge then current. At that time the category of form of motion could only be employed to explain the simplest forms: those of the inorganic world. Only when it was possible to solve the problem of the passage from inorganic to organic nature—and when knowledge in that area had sufficiently progressed—could the category of form of motion also be applied to the organic world[13].

But there was another aspect of the problems raised by the state of the sciences at that time that affected Engels's theoretical model. If, for the time being, the category of form of motion could only be applied to inorganic matter, in its abstract formulation it also reflected the limitations of the conceptual horizon of that time.

This category had, moreover, to be connoted by an entire grammar of scientific knowledge. It was possible to predict certain lines along which that knowledge would develop and the new importance of these lines; but unless all this was translated into results that were also conceptually appreciable, it was necessary to stick to the terms of that grammar. And it was, indeed, a grammar basically capable of conjugating the states and changes of state of matter, beginning to come to grips with new notions like evolution and adaptation but, perforce, only in general and schematic terms.

So that it was the category of form of motion that was to remain connoted by theoretical limits. And indeed in its applicability, "historically" possible, to the phenomena of the inorganic world Engels was able to foresee properties and logical attributes belonging to it that were obviously still wedded to the categories remaining in force.

So the problem could not be solved in mere terms of historical continuity. It was not enough to expect that the progress of science would yield results capable of repairing the breaks in that continuity in such a way as to enable a complete picture of it to be given, together with a full historical reading in the terms conceived by Engels. In fairly recent times, as that point was approached it was accompanied by a profound shift in the conceptual horizon. The implications for theory enabled the establishment of nexuses of quite different importance.

These nexuses shifted the problem of continuity onto a plane strikingly different from that of the Engelsian conception. In fact, at the very moment when it seemed that those "processes of linkage" which would complete the sequence of that "continuity" could be achieved, the type of linkage that became possible assumed a significance far beyond that of pure and simple "process of linkage": broader connexions were implied, together with different presuppositions of the unity of reality; it made possible the establishment of nexuses of quite other importance, capable of justifying a unitary conception of reality resting on quite other foundations.

In other words, the progress of science after Engels was, in a sense, to confirm the necessity he had indicated and asserted to be plausible—the necessity of finding the moments and processes of linkage that would reveal the shortcomings of the "hard and fast lines. . . . incompatible with the theory of evolution"[14].

This development of science would effectively succeed in establishing new connexions by opening the way to new areas of research and propounding fresh problems. But all this would involve—as in fact it did—a profound change in the grammar of scientific knowledge. And this was the logical consequence not only of progress but of the advent of new sciences and the change in the relations between the most recently developed sciences and those already long established.

The new domains of the sciences of the organic were to cause a shift of the entire conceptual horizon. So that the solution of essential problems, such as the passage from inorganic to organic, was close to being achieved, the labyrinths of the Engelsian "prehistory" of human thought would be explored; but all this would not be resolved simply by the insertion of missing pieces into a mosaic of continuity as conceived according to the Engelsian model.

If still largely incomplete, the pattern of the mosaic is now more evident, but our points of view have considerably altered. In the previous chapter we noted the significance of the most striking findings of recent science and where they pointed, and we mentioned the new terms in which the problem of a unitary conception of reality is now propounded. Here we shall confine ourselves to emphasizing how the most important point in the change of perspective is, without doubt, the possibility of establishing a systematic nexus between nature and knowledge at the various stages and levels of the evolutionary process.

On this basis a continuity in the man-nature relationship should be sought not in a linear historical development, observed from the outside, but rather in the very foundations of a unitary conception of reality. In other words, that continuity—in the new perspective—emerges as based on a systematicity of nexus which is inherent in the very process of evolution and therefore does not require to be assumed or derived, either by means of analogy or on the basis of a simple correspondence.

The historical conception is thus not superimposed on the world of nature or extended to it on an analogy between the laws of thought and the laws of nature: the laws of thought are not applied to those of nature, since thought is a product of nature and by analogy corresponds to it.

In effect, a purely diachronic sequential continuity, in which each level of evolution took its proper place, in order with respect to the others, so as to produce a continuous line running from inorganic nature to man; a linear unfolding in history with the gaps all filled by the results of experimental research:—such a continuity could, from a theoretical point of view, quite easily be "interrupted" afresh and cast in doubt every time we wanted to introduce distinctive characters and criteria of differentiation between the principal levels of reality (inorganic nature, organic world and human thought), in order to predicate the characters most peculiar to each of these. This was, in substance, what took place in the criticism that followed on the *Dialectic of Nature*.

Even if one accepted the Engelsian perspective, according to which science would sooner or later demonstrate the continuities (most importantly, that between inorganic nature and organic world), there was nothing to stop the differentiations assuming the force of categories—on the basis of distinctions that continued to be possible.

As conceived by Engels, this continuity of the man-nature relationship remained open to grave doubt and on more than one occasion was to reveal its vulnerability.

Leaving aside other criticisms from standpoints very far from the Engelsian one, the continuity was contested within the Marxist camp just as we have mentioned above. Lukács's early position—while still under the influence of vitalism and German historicism—made a clear distinction between natural and socio-historical sciences, both at the level of content and that of method (analytical in the first case, dialectical in the second)[15]; but it is also well known that even later, when he returned to the problem, he made a neat separation between inorganic reality and organic and social reality[16].

According to Lukács, objects and events in inorganic nature are characterized by synchronic microstructures and diachronic macrostructures, and are incommensurable if related to orders of greatness appreciable in the human conspectus. They possess, so to speak, "more intrinsic dimensions", such as to appear rather as repetition and cycle, and thus cannot be described in terms of diachronic unfolding. So that they are, in effect, excluded from a "historical-documentary reading and thus from dialectical treatment"[17].

Quite clearly the problem is posed with reference to a diachronic-sequential continuity (that of Engels, indeed) which can only be read in a historical-constatative way (or, in Lukácsian phrase, historical-documentary). So that continuity is subject to doubt, since, by the way it is constituted, it admits differentiations that assume the force of categories.

Another equally significant criticism of the type of continuity implied by the Engelsian model is represented by Gramsci's position. In this case the problem is propounded with reference to the relation between the history of nature and the history of man, and extends beyond categoric differentiation.

For Gramsci the history of man must be not only clearly differentiated but also located on another plane, in the position of prime importance. It cannot be understood simply as a part-unfolding of the more general history of nature: "man does not enter into relationship with nature in a simple way, by the fact of being, himself, nature, but does so actively, by means of work and technics"[18]. Indeed, a history of nature exists only in so far as it can be subsumed into the history of man. In fact, "if all nature is history, it is so by virtue of human activity, economic activity, scientific activity and others"[19].

And the part of that activity which is sometimes mistakenly referred to "human nature" should instead be referred to "the totality of social relationships which determines a historically defined consciousness"[20].

For Gramsci the very concept of matter must be historically assumed—but only in the sense of the history of social relations of production. "The various physical properties (chemical, mechanical, and so on) of matter which together constitute matter itself. . . . are taken into consideration, but only in so far as they become an "economic element", an element of production. Thus matter must not be considered as such, but as socially and historically organized for production; and thus natural science must be conceived as essentially a historical category, a human relationship"[21].

So the Engelsian continuity laid itself open to question from two standpoints: the introduction of differences having the character of categories between the principal levels of reality; and the reaffirmation of the primacy of human history in the subsumption of natural history.

Now there is a first order of considerations to be made with regard to the categoric value assumed by differentiations introduced into the continuity of the man-nature relationship. These considerations refer, more precisely, to positions like those adopted by Lukács in his last phase, yet they are also of significance for the problem as a whole.

They have to do with quantitative-qualitative differences in the dimensions of space and time.

With regard to objects and events in the inorganic world, if it be true that in the scale of the human conspectus these tend to appear as cycle and repetition, we cannot neglect the fact that technical and scientific progress have equipped that "human conspectus" with instruments (electron microscopes and telescopes, large accelerators, etc.) and, above all, with research methods so powerful that the synchronic microstructures and diachronic macrostructures tend to become less and less inaccessible.

Merely by way of example, suffice it to remember that the discovery of a galaxy more than fourteen billion and a half light-years distant from us makes it possible for us to look back over the same distance in time and witness the formation of a galaxy.

So that the exclusion of such objects and events from any idea of unfolding cannot aspire to become a distinction of the categorical type. Just as a categorical distinction cannot be introduced between organic and human worlds on the basis of the necessity-possibility alternative: when, instead, this alternative must rather be referred to our various models (macroscopic and microscopic) for interpreting reality[22].

As long as it is a question of categorical differentiations concerning the historical conceivability of inorganic, organic and human worlds (for which this conceivability is either excluded for the first or admitted for the first and second, only in as much as it is subsumed by the third), the problem will substantially consist in a difference of dimensions referrable to the evolutionary scale itself and/or to different models of interpretation.

But both in the viewpoint of the later Lukács, which excludes inorganic nature from historical conceivability, and in that of Gramsci, which admits a historical conceivability of nature only by subsuming it into the history of man, not only is the debate on continuity implied in the Engelsian conception of natural and human history resumed and a predominantly historical-documentary reading of that continuity seen to be unsatisfactory; together with this, another problem of a gnoseological kind emerges, with respect to which the foundations of the continuity also appear insufficient.

For both Gramsci and Lukács the cognitive process was to be distinguished from natural history and placed over against it. This was possible because the continuity implied in the Engelsian formulation did not admit any type of conception other than the historical-documentary: so that human knowledge had either to be understood as the sole subject of that historical conceivability, in such a way that the history in which it could be inserted (i.e. human history) would constitute the only theatre in which it unfolded (as was the case with the human history-natural history relation in Gramsci); or it was necessary to exclude, as object of what could be historically conceived, from human knowledge everything that could not be referred back to it in terms of that same conceivability (as was the case with Lukács with the exclusion of historical conceivability from the inorganic world).

In order to formulate those problems differently, it was necessary for scientific research to develop far beyond the terms of reference that were available to Engels, but also to the next stage of Marxist thought, that of Gramsci and Lukács. In the end it was necessary to arrive at the real breakthrough that the most recent findings have brought about. In connexion with which there can be no doubt but that the notion itself of continuity has been shifted onto another plane and can thus be given a different basis.

The problems posed by critical reflection subsequent to Engels well demonstrate which difficulties can now be resolved and what terms they can be reformulated.

Already from the XIX century onwards, the changes in the notions of subject and object of knowledge directly reflected the developments of science, the modifications in the scientific system and the relations that characterized its organization. Direct references to this can already be found in Engels's thought. They were to become ever more cogent in the subsequent development of science, to the point where the notions of subject and object were redefined in the ambit of scientific research itself, both with respect to the problems directly implicated and, in certain cases, with respect to the specific research topics themselves.

In this perspective stands the attempt on the part of genetic epistemology to furnish a natural foundation for the knowing subject. Nor, obviously, is it any longer a question of unitary and apodictic subject, but rather of a complex interweaving of psychological, biological and other processes.

Moreover, this line of research unfolds in a scientific context characterized by a multiplicity of levels and sectors of research all tending towards closing the traditional gap in the relation between biology and knowledge. Research, therefore, not only avails itself of the findings and references coming from those different levels of investigation, for the study of subject-object interaction in knowledge, but it also sends us back to that plurality of levels in the redefining both of knowledge itself and of the "decentered" notion of its subject.

The salient features of a perspective of research of this kind thus become the following: the structural circularity between the various levels of reality, the configuration of knowledge in terms of process and the integration of the subject in the historico-natural process (natural history of the subject and general history of nature)[23].

With respect to the traditional conceptions of subject and object, genetic epistemology opens anew the debate on the postulates both of pure objectivity and of pure subjectivity "in favor of a postulate of subject-object interaction that develops genetically"[24].

The conception is no longer one of knowledge and a subject of knowledge that uncovers what was formerly hidden (with the resulting distinction of subject and object of discovery in terms of traditional logic). Knowledge is conceived, instead, as a new working-out of what does not yet exist, by virtue of which the subject-object relation is a genuine interaction: what were traditionally defined and distinguished as subject and object constitute the terms of an interaction that develops genetically in the growth of knowledge as a process of construction.

The new perspective, which began to be outlined in the nineteen-sixties, thus consists in a shift from an epistemology of representation to one of construction. Certain implications of this shift have been considered in the previous chapter. Here we must stress the radical changes it wrought in the relation between knowledge and reality: this relation no longer consists of a correspondence given in terms of figurative representation, but rather in an adaptation in a functional sense[25].

The new perspective that emerged in the nineteen-sixties, in both the practical and theoretical areas, showed the constructive character of the limits of human knowledge, of its finiteness. Its preconditions and limits, as well as its obstacles and conditionings (which were to be "purged" from a real "pure" knowledge) were viewed as something constructive, as irreducible matrices of knowledge itself (real procedures of knowledge, or better, its historical conditions). As these conditions and real procedures of knowledge were further identified and investigated, that purging operation that traditional epistemology had performed on them was once more called in question.

So that in this new conspectus the traditional reference to a "self-centered and all-knowing" subject, whose cognitive capacity suffers no limitations or conditionings, no longer holds. Such a subject would be alien from the world which it nevertheless tried to describe. What traditional epistemology had cut out—i.e. the real historical procedures of knowledge, the "fabric" of knowledge—is reintroduced; and thus we can see how the subject of knowledge was actually concealed till then, expelled from the world, estranged from it.

The subject is finally made to belong in the world once more: its description of the world becomes its description of itself as it describes the world[26].

But, what is more important and connected with this shift in perspective is the whole development of cognitive sciences, psychological and biological sciences and sciences of systems. It is these results that have combined more and more to redefine the nature and limits of knowledge, along the lines of a "natural epistemology"[27].

Knowledge is no longer assumed *ab initio*, or as an ideal unquestionable starting-point. The activity of knowledge, freed of the automatic reference to an abstract subject, is reconstructed in reference to biological and psychological conditions[28] or, as I should prefer to call them, biopsychic conditions.

In this change in perspective an important part has been played by the contributions from "second cybernetics" and the reworking of concepts—in self-referential terms—in the biological sciences, in organization theory and in systems theory[29]. Key concepts—in the sciences of man as in the natural sciences—like the concepts of information, structure, organization, possibility, order and so on—no longer labour under the necessity of retaining a basic observation point[30]. Thus it has been possible to redefine them in a way that transcends the traditional epistemological picture. It appears feasible to develop a theory of the observer by doing away with that same observer as an external condition of the domain of observation.

The result was, perforce, that a multiplicity of points of view was recognized; a multiplicity that concerned the modalities of consitution of the cognitive universe and of the identity of a subject.

Any image of the subject one might nowadays propose would have to be polysystemic; the standpoint of that subject would have to be defined in a rejection of the search for a basic observation point; and that standpoint would hence rather be one where systems endowed with diverse logics and histories encounter and clash with one another[31].

So that now, the failure of a conception of subject (and, indeed, the collapse of the problem of subject) according to the canons of traditional epistemology, enables us to return to the problem of knowledge in the context of a continuity of the man-nature relationship differently understood. What breaks down, in sum, is an anthropocentrism, abstractly conceived and abstractly founded (or unfounded!).

Thus it becomes possible once again to reinscribe the cognitive process in the natural process of evolution, without making human thought the sole subject or the isolated subject of that cognitive process. And so, in theory as in fact, the categorical distinctions and exclusions of everything that cannot be appreciated unless referred to a human conspectus break down. Not only does this conspectus enlarge and expand, but its focal point can and must be shifted towards different dimensions from that of relation to a merely human scale (a scale that turns out to be wholly "relativized" by the developments of theory we are dealing with).

But in this way historical anthropocentrism, too, begins to break down. An entirely subjective conception of human history begins to totter, and with it the conception of its relationship with the history of nature. If human thought cannot be isolated in the cognitive process, one begins to understand how history, that same history of man, is not only such, nor *a fortiori* that of man as the sole subject of history.

If the cognitive processes are inherent in the self-regulation of the systems of evolutionary mechanisms, then they perform functions of exchange and interrelation; so that they cannot be represented in subjective terms, that is to say, they cannot be referred back to a subject of knowledge as being the source of that knowledge.

By the same token, everything that has been traditionally thought of as object, merely susceptible of being known—in a cognitive process, differently understood, as a system of interrelations—turns out to be interacting, itself too a factor in the cognitive process, instead of a mere passive datum.

So that we cannot advance any distinction-separation that will isolate an objective reality, conceiving it as more or less penetrable and, in any case, as an inert or residual *donné* of whatever knowledge is in question; for, whatever the case, it will correspond to that degree and possibility of knowledge in which it is comprehended: comprehended, that is, as an active part of the cognitive process corresponding to a determinate degree of evolution.

And if, in the same fashion, what is traditionally indicated as the subject of the cognitive process is "comprehended" in that process—comprehended in the sense that it cannot be isolated since it is an active moment in that process—we shall easily see how the cognitive process can by no means be excluded or separated off from the natural process of evolution; rather, it is fully inscribed in that process.

In other words, the cognitive process does not lie outside or over against nature as being referred to a knowing subject; nor does it exclude from itself, or distinguish as alien, nature as an object of knowledge, a mere passive or residual datum. We can no longer deal with knowledge in extrinsic terms of reference, nor in terms of subject and object, such as would bar it off from the evolutionary process; the cognitive process and the evolutionary process move together as one.

But as far as the problem here is concerned—the problem of the continuity of the man-nature relationship in a unitary conception of reality—to transcend such a traditional conspectus and its logic means to free the conception of the first term from all subjectivist residue and that of the second from all objectivist residue, as regards the problem of knowledge. So that knowledge belongs to man inasmuch as it belongs to nature itself, and the progress of the one is accompanied by the evolution of the other.

Thus it will be seen how a different and genuine foundation for a unitary conception of reality is given by the possibility of fully rethinking knowledge with nature in the original nexus, and by the fact that the nexus—through the synthesis of form—is expressed at all degrees and levels of the evolutionary process.

Hence, from the "most intrinsic dimensions" of objects and events in inorganic nature (so elusive to perception in the human conspectus) to consciousness—by which human thought and will unfold, at each of these levels, as in all others—nature and knowledge are welded into one, comprehended in each other.

Form, i.e. the active synthesis in which this comprehension is expressed, is certainly something different and corresponds to the diversity of evolutionary levels. It will be more closed, the more "implicit" is the connexion between the two terms in inorganic nature. It will be more and more open in the case of living matter. Till we reach the most open form of all—which might be defined as "exposed"—that of man's conscious thought, in which the unfolding of that connexion is wholly "explicit".

We have seen elsewhere how the nature-knowledge nexus is expressed in the synthesis of form on the basis of the property of that synthesis, by virtue of which form is both an understanding and a performing at one and the same time. Now, the variety of degrees of explicitness of that nexus and thus the differing character of form, from more closed to more open, is due to that very correspondence of form to the different degrees and levels of the evolutionary process.

So that knowledge is connected with nature in every form, even in the most closed ones and in those where the nexus remains entirely implicit,—inasmuch as form is the comprehension of that towards which it tends in its becoming; and, since it already contains what it tends towards, the self-comprehending coincides with the self-constructing.

The correspondence of the synthesis of form to different degrees and levels in the evolutionary process is not a correspondence to levels considered as steps or passages that are somehow necessary and can be presupposed as "goals" in the evolutionary process. Those levels cannot be thought of as detached from the forms that concretize them, neither as prior nor subsequent to those forms, as their aim or result. Rather can they be represented as *tendency,* as that towards which form tends; but since form already contains that towards which it tends, we may aver that the tendency is nothing less than the form's tending towards itself.

Now it is this very characteristic of tendency in form and its correspondence to the various levels of reality that make it possible to conceive of the dual connotation of the nexus—cognitive and also natural—in accordance with a tension inherent in the form itself (it will help to bear in mind the "psychoid" nature of this tension).

The systematicity of the nature-knowledge nexus, at the various degrees and levels of the evolutionary process, can be understood as "construction" or, better still, as a tending towards construction of those same levels. A conception of this kind moves in a direction exactly opposite to that of evolution proceeding by "deducible" derivations. For if that were the case, the correspondence of the nature-knowledge nexus to the various degrees and levels of the evolutionary process would either not exist at all or could only do so in deterministic terms.

Form, on the other hand, as active synthesis of the evolutionary process at every degree and level, corresponds to these only in its capacity as construction—tendency. Itself is indeed nothing other than tendency (or tendency towards itself).

Furthermore, since the various levels to which form corresponds cannot be determined *per se,* outside form, and since this correspondence is non other than a tending, and tendency is form itself, it follows that the synthesis of form represents events and situations which at any one time are particular or relative. The general process of evolution—as it can be configured with respect to the form that constitutes its active synthesis, at every degree and level—thus appears as a continual approximation towards innumerable (relative) situations.

The whole process of evolution, in its totality, resembles a kind of continuous, generalized experiment. Thus it is the nature-knowledge nexus itself, or its correspondence to the various degrees and levels of the evolutionary process, that proceeds experimentally.

In substance, if it is true that, in evolving, nature knows itself, we can say that it does so experimentally; as occurs at the highest stage of the process, in the case of human knowledge.

From this point of view we can say that active synthesis as constituted by form is the abstract formulation, or rather generalization, of the mode of procedure of evolutionary experimentation in the connexion between nature and knowledge.

Form supplies a paradigm for the unified conception of reality. As we consider the above-mentioned process of approximation and experiment, this conception enters as an exigency within scientific research itself and redirects the findings of that research in the same measure as its adoption is made possible and necessary by those findings.

Ultimately form, as active synthesis in the process of evolution, enables us to repropound the unified conception of reality, on the basis of the systematic nexus between nature and knowledge that is expressed in form. For the nature-knowledge nexus in intrinsic terms establishes the continuity implied in that conception.

It decisively shifts the perspective within which the gnoseological problems relative to that conception have traditionally been framed; and it enables us to move beyond the categorical distinctions that have been proposed and reproposed by separating the principal levels of reality from one another.

Perfectly consistent, moreover, with the non-deterministic systematicity of that nexus is its character—not merely specific[32], but tendential and relative to the degrees and levels of evolution to which the synthesis of form corresponds. Hence the configuration of the evolutionary process as a continual approximation, through the multiplicity of particular events and situations that form each time represents.

The properties of its unfolding do not, therefore, present categorical differences such as to require a distinction-separation of the socio-historical level from the natural level along the lines of the traditional differentiation; rather do these properties enable us to refer to a type of relationship of particular to general and local to global, that is the same in the scientific analysis of the one evolutionary level as it is in the analysis of the other.

Form, in sum, interprets the results of research in the area of natural science according to the same epistemological connotations as those of the socio-historical sciences. In the one domain as in the other, it shows the properties of an identical singularity.[33]

The substantial alterations in the grammar of scientific knowledge obviously arise from the new dominion of biology, as well as from the striking progress made by physics in this century—and this not by chance. We have witnessed the establishment of a different relationship between local and global in the scientific intelligibility of phenomena[34].

For all these essential requirements, including the last named—which has an epistemological character and can be derived from the others—form constitutes a real scientific paradigm[35], enabling us to confront the task of reconnecting the socio-historical sciences with the natural sciences.

Their original link was broken with the establishment of the Galilean paradigm[36]. By creating a local-global relationship that was actually inverted as compared to the previous one, this paradigm influenced an entire scientific typology, mechanico-physical above all, that aimed at quantitative prediction[37].

Starting from the establishment of the scientific paradigm that was founded on Galilean and Newtonian physics and went on to permeate the whole subsequent development of modern science, the relationship of general to particular has tended to affirm itself in such a way that those sciences aiming in the direction of the general have tended to demote the particular, even to the point of neglecting it; in this way they acquired an ability to make quantitative predictions and assumed a strong scientific status. Whereas those sciences for which the particular retained great importance were destined—according to the paradigm—to resign themselves to a weak scientific status on pain of achieving poor results otherwise[38].

Hence there occurred a real crisis of delegitimization of the human sciences[39], together with a neat, categorical distinction between them and the natural (above all, mechanicophysical) sciences; the end result was that the scientificity of the latter came to be thought of as paradigmatic.

Obviously this turnabout had to do not only with a paradigm of scientific knowledge: the adoption of a definite scientific perspective could not but be accompanied by radical changes in the conception of man and his relation with nature.

For these reasons the adoption of a new scientific paradigm like that of form, and the relinkage of the socio-historical sciences with the natural sciences which that paradigm now enables us to effect, take on a much more general scientific and theoretical significance.

We have here not merely a reconjunction made possible in terms of epistemology: alongside this, and indeed as its fundamental premiss, stands the coherent totality of all the properties and values represented by the paradigm of form. On the one hand this totality makes that reconjunction an effective possibility, while on the other it reinscribes it in the perspective of a unified conception of reality in new terms (such as those indicated by the original nexus of nature and knowledge).

The relinkage between natural and socio-historial sciences can be seen in terms of a genuine refoundation as against the traditional separation of their respective domains. It necessitates a redrawing of the boundaries of research areas and a rethinking of their significance. Actually it points us towards a new and more radical conception of the relation between natural and human evolution.

In the light of those implications for theory and practice, the paradigm of form appears capable of satisfying the necessary and sufficient conditions for resuming the problem originally propounded by Engels: to

bring "the science of society, that is, the sum-total of the so-called historical and philosophical sciences, into harmony with the materialist
foundation" and reconstruct it thereupon[40].

But the very extent of that reconstruction—when we bear in mind the
new terms now available for a unified conception of reality—requires
that the materialist basis itself be refounded. The character of the new
scientific paradigm represented by form can be gauged in this very possibility of refounding. In effect, this possibility would appear to offer a
new materialism (modern materialism) only if, and in so far as, it is
transposed to within the sciences[41].

The new materialism permeates history and the sciences of man inasmuch as it relinks them in scientific areas that no longer admit the traditional distinctions and separations between the principal levels of
reality. The reconjunction thus gives them a quite new importance. They
take on a further solidity and extension.

To sum up, a continuity in the man-nature relationship is established,
in differently based terms. As we have seen, the basis is the systematic
nature-knowledge nexus in the process of evolution.

Knowledge, indeed,—all knowledge and not that merely of man as
sole subject of knowledge—rather than returning to its crude materialist
basis invokes the depth of a constituting materialism.

It is no longer a question of a materialist basis from which the laws of
man's thought and action are deterministically derived, nor of an extension, pure and simple, of these laws to the world of nature. The very
man-nature dualism—in every one of its formulations of subject-object
distinction, in the cognitive process and as regards the process of natural
evolution—is completely overridden, and thus requires no hypostasis
nor primacy, either subjectivist or objectivist.

Today we can effectively conceive a continuity in the man-nature
relationship, repropounded in new terms, whose significance is historical and embraces the refounding of our history. And this history can
be refounded on a materialist basis, but not in a limited, one-sided fashion; rather in the perspective of a unified conception of reality.

The new materialism is a necessity that makes itself felt within the
sciences as an essential part of their constitution. It enables the transcendance of many of the limitations of the Engelsian scheme and of the
several critical qualifications to which it has been subjected by Marxist
thinkers themselves.

Gramsci's reservations on certain of Engels's formulations in *Antidühring* are well known. In his opinion, these formulations could lead to
"deviations" of the schematic materialist kind like those of Bucharin[42].

And in the same passage Gramsci expressed his fears of an idealism arising out of a quite opposite conception like that he fancied he could detect in the early Lukács.

But especially in view of the dogmatism that held sway in the USSR in the period of Stalin and the schematic definition of dialectical materialism that derived therefrom, certain Marxists, then and later on, reserved their most serious criticisms for "the attempt to *deduce* particular dependences from general schemes of interdependence": here they saw the greatest danger of schematism and a resulting hardening into dogma[43].

It is important, therefore, to stress that the conclusions we have reached how on the reconstruction of the materialist basis of history and the sciences of man should be understood, explicitly rule out a deductive relationship, however interpreted. Actually, the entire perspective, indicated by the paradigm of form, *at the general level* excludes any procedure by derivation or deduction.

On the contrary, that paradigm appears to offer a quite different strategy in relating particular to general: one capable of reflectively reconjoining the properties and connotations of history and the sciences of man with those of nature, and of directing our focus on a process of evolution, natural and human, that emerges as a continual approximation through the multitude of events and particular situations represented by form. The process, in short, runs in quite the opposite direction to that derived from a dogmatic and deductivist interpretation of dialectical materialism.

Indeed, that process convalidates those characteristics of specificity, and historical determination—in a word, particularity—that have been advanced in defence of the non-schematic, non-dogmatic character of the Engelsian *Dialectic of Nature*[44].

To sum up, it is precisely by reappraising and elaborating the most valid and relevant parts of Engels's method dealt with in these pages— the part bearing on the relationship with the sciences—that we can avoid schematism in our treatment of his thought.

By the same token, direct comparison with the latest findings in the sciences (natural and socio-historical) enables us to reflect on the continuity of the man-nature relationship—reformulated in the new terms necessitated and made possible by those findings. In this way we can once more tackle the problem propounded by Engels, of a relinkage between history and the sciences of man and their materialist basis.

This leads, I repeat, in the direction of a modern materialism that takes its place within science. And the new scientific paradigm, which may preside over that relinkage, finds its verification in its demonstrated ability to favor the relinkage.

As a result, I believe, all deductivist dogmatism is out of the question, together with any superimposition of the new paradigm on real cognitive processes. On the contrary, a valid interpretation of the theoretical implications of important scientific findings may bring considerable advances in knowledge as and when it becomes able to establish itself as a necessity inherent in the development of science for the reformulation of problems.

As has been rightly pointed out, this "need for a theoretical dimension within science" is the path the sciences have taken "carrying forward, in our century, the theoretical revolution that, in the XIX century, had already transformed the biological and social sciences"[45].

Thus liberated from the stranglehold of deductivist schematization and dogmatism, the reconstruction of the materialist basis of history and the sciences of man can take a clearer shape. In actual fact, it is that materialist basis that is reconstructed. If referred to the continuity of the man-nature relationship, it must be considered as open on both sides of that continuity. Just as the scientific paradigm corresponding to the unified conception of reality is open on both sides. Our reconstruction, therefore, will lead neither to a materialist history of society alone nor to a sort of social naturalism. What it will instead favor is a more comprehensive notion of man himself and his relation with nature, as well as a different mode of understanding knowledge within that relation.

On this basis a different materialist conception of history and of the sciences of man takes shape: in the dual significance of the ampler area, both scientific and theoretical, in which that history is contained, and the historical problem (of historical praxis) of the relationship with nature.

As regards the first point, it is obvious that if the materialist basis on which history and the human sciences can be "reconstructed" is redefined vis-a-vis the twin areas of human and natural sciences and their relinkage, that ampler area mentioned above can no longer be marked out according to a traditional distinction-separation grid. It must be viewed in a quite different perspective: the very notion of man will be re-embraced in a re-established continuity of the relationship with nature and this in the new terms of reference of the systematic nature-knowledge nexus.

In the new terms of reference, man is, as it were, immersed at a deeper level of his comprehension, both subjectively and objectively; indeed, this distinction tends to lose its meaning—or, to put it another way, the relation between the two levels becomes one of increasing reciprocity.

By the same token, at deeper levels, the very notion of man will be affected by relationships extending far beyond the traditional limits of that notion, beyond the range in which it can be recognized; where it is evident that this expansion of the range of recognizability is linked with that of man's cognitive (subjective-objective) potentiality.

And when we consider each new relationship that can be established in this sense, what does it ultimately represent if not a new acquisition of knowledge, or better, an enrichment and reinforcement of the knowledge of man, both in the subjective and objective senses?

This knowledge obviously tends of itself to comprehend the other and to relate itself to it. But this does not take place in a one-sided relationship of pure and simple assimilation. It occurs rather through a mechanism of recognizing-relating, through the tension of a relation that now bursts its former limits.

The act of cognition can therefore be configured as a prolonging of itself, but, be it added, a prolonging and projecting of itself towards the other by itself, along the lines of a new, recognizable tension: according to a comprehension that already contains that towards which it tends, which is to say, through a comprehending and performing, together, that belong to the synthesis of form and its recognizability.

It will thus be perfectly clear what is meant by transcending the limits of reference to man as mere subject of knowledge or isolable subject of knowledge.

Man proceeds towards a self that is ever more comprehensive and ever less isolable (i.e. towards a comprehension of self ever more expansive and less univocal). Which is equivalent to a deepening of his knowledge; whereas the subjective-objective distinction tends increasingly to become meaningless.

Moreover, to outline in these terms that deepening of knowledge well demonstrates the advantage accruing from a change of perspective like the one which takes account of the systematic nexus between nature and knowledge.

In this way, a shift away from the traditional "point of origin" in the cognitive process announces itself (away from the notion of an "origin" that can be distinguished and isolated and stands at the beginning of a process, serving as the "source" of that process).

Instead, this shift involves a transposition to the inside of the cognitive process of each of its terms of reference (which are comprehended in the synthesis of form, i.e. in a comprehending that is also, at the same time, a performing).

This shift to within the cognitive process of each of its terms of reference, including that of a supposed point of origin (represented by the traditional view of man as a mere abstract subject of knowledge), which is recomprehended by the process itself, in the sense of a complete belonging to it, brings about a reinforcement of knowledge. Knowledge acquires new values and new properties.

Firstly, a close relationship is set up between each point of view, will or cognitive intention and that towards which that intention tends (and which can no longer be defined or isolated as a mere object of knowledge). If we would employ a logical distinction, transcended in the facts themselves, we could say that the knower belongs to the known.

Secondly, the transposition of each of the terms of reference to within the cognitive process involves an ability to recognize the very pattern of unfolding of that process at the moment at which it is operative (and this corresponds to the recognizability itself of form, according to which it operates). This ability (internal) to recognize can be understood as a sort of ''second view'', or knowledge of knowledge: and, as we can see, this constitutes the only effective knowledge in terms of pure knowledge.

Thirdly—with the multiplication of the effect of reciprocity in the subject-object relationship, as knowledge proceeds—the transposition of the cognitive point of view to the inside of the process enables the perspective to be, as it were, doubled; this as a result of an inverse projection which is always possible. Thus we have a ''reciprocated'' perspective, a dual perspective, transferred and re-opened onto the other view; its duality indicating not an objective standing opposite a subject, but simply one ''mode'' of knowledge. And in this mode of knowledge man can draw on the elements of his own knowledge; for he is relinked to that mode, comprehended by it, recognizes it and recognizes himself in its form and tendency. Thus he knows his own knowledge and it knows him.

So we see that there are three requisites or properties belonging to a cognitive process seen in this changed perspective—or should we say, connected with an acquisition of new capacities on the part of that process, by which it is reinforced?

This knowledge, that finds all its points of reference within the cognitive process itself, which is intimately correlated with the evolutionary process itself, thus ultimately comes to correspond to those quite new

problems of knowledge that arise out of a scientific research trespassing
beyond the bounds of the traditionally knowable: and this is so whether
we propose to analyze scientifically the cognitive functions themselves,
whether we attempt to perceive the "reason" behind matter, or whether
we try to interfere with the results of evolution.

The further we press the man-nature relationship in a reformulation of
the problem of knowledge in such a way as to exclude all one-sidedness
from that relationship, the more powerful will our knowledge become;
with regard to its terms of reference, this reinforcement will be not
merely useful but also necessary, when we confront the problems and
implications arising out of the most significant findings of science.

It cannot escape notice how the one problem, that of knowledge, links
itself to the other, the historical one of the relationship with nature.

And as regards historical knowledge in the strict sense—which I shall
treat more fully in the next chapter—it is also quite obvious that our new
perspective does away with the "humanistic" prejudice and precludes
our clinging on to or reformulating a conception of history in terms of a
mere dominion of man over nature. The field must be weeded of a prejudice that hinders any attempt to consider relations occurring outside the
sphere in which human action and thought have traditionally been located and confined; a prejudice that also obstructs attempts to modify the
one-sided abstraction of the man-nature relationship conceived in terms
of mere dominion.

In the last analysis, the cognitive and theoretical problem of the man-
nature relationship ultimately corresponds to the historical and practical
problem of that relationship.

If for a long time "the evolution of life was regulated by retroactive
interactions between the structures of life (with a greater degree of order)
and the natural environment (with a lesser degree of order) through ac-
commodations that spontaneously became general, today the problems of
life and its perpetuation tend to become a question of historical choices
involving the relations between people. The responsibility for evolution
is no longer entrusted to the instinctive defence mechanisms with which
the individual is equipped (and which constitute the almost exclusive
source of current value judgements); it has now become a problem propounded in terms of historical choices, involving the entire species"[46].

But we are dealing with an age-old historical problem that has been
propounded and repropounded at different times and stages in mankind's
social evolution. Already in traditional social formation[47], with respect
to our animal origins, "the insufficiency of the equilibrium achieved by

the species with respect to the historical utilization of the forces of production'' was clear, and the need was recognized for a "deeper equilibrium that can be reached at the level of conscious regulation''[48].

Man's whole natural historical past took the form of a sort of experimental history of the species, urging itself on the then world to be rein-
terpreted and reoriented, drawing on a deeper level of man's
consciousness and capacity for regulation.

This need to move towards conscious regulation or, if we wish, to a
more open expression, at the level of historical praxis, of what was a
tension within the species, became more impelling in the formation of
modern society. "The awareness of the genesis of man, of his unstable
equilibrium with nature, of his irrational appropriation of nature unto
himself, of the interruption in the harmony and the need to reconstruct
it''[49], became awareness of an objective task that must be performed if
the species was to survive. The problem posed itself, therefore, in col-
lective terms of historical choices involving the whole species.

The problem posed by Engels—how to reconcile the science of society with the materialist basis so as to reconstruct the former upon the
latter—ultimately had the same significance, and required, on the one
hand, a radical revision of the presuppositions and methods of the sciences, and, on the other, a different, more direct social use of these. This
also involved freeing scientific knowledge from such of its characteristics, bearing on the division of labor, that obscured a clear view of the
situation and the perils it held[50].

Now, more than a century after that analysis and faced with the problems of industrial society and technological progress in our world, we
can feel that historical problem—which even then appeared inescap-
able—as a matter of dramatic urgency.

In our present state we find ourselves at a point of no return. For the
first time in history man's most elementary needs, those most nearly
connected with his animal nature, may have become unattainable.

The fact is that a more and more general crisis is assuming the
character of irreversibility. Not only are elementary necessities sacrificed
to other artificial and gratuitous necessities, restoring mankind to a
one-dimensional state and mutilating him in the process, but also those
natural resources that should satisfy mankind's need are currently threatened. We are witnessing new forms of plunder that strike with an absolute finality at the sources of primary necessities, through a rarefaction
of the objective presence of use values that are among the most essential
for the whole species, such as breathable air, drinkable water, vegeta-
tion, the renewal of the primordial forms of nature[51].

With the way we are pursuing economic and technological progress at present, more and more of nature is being subjected to human transformation in an uncontrolled, destructive fashion; man's basic and necessary relationship with a friendly nature is thus relegated and threatens to disappear altogether. The outcome of this uncontrolled and alienating expansion will be that nature, thus transformed, will come to seem something monstrous and hostile that we can no longer exorcise. And even the elements with which we substitute it (planning, comfort, design), introduced into the artificial environment that replaces our interactive relationship with nature, having no real basis on which to stand, will lose every function, every symbolic property of support. We shall be left with them as the mere space-consuming leftovers of our alienation.

But it goes further than that: following this logic, the powers that man has pursued and achieved over his own species—from the ability to alter the results of his own evolution, to the capacity to destroy himself—are so frighteningly one-dimensional that he has little idea of how to employ them. Thus we come bump up against the most striking, most paradoxical expression of an absurd solipsism.

At this point we need a radical historical conversion—both theoretical and practical—able to reflect that relinkage of man with nature, as required by the progress of science, i.e. by the new problems of knowledge and the praxis connected with them.

And note that this is a *conversion* also in the specific historical sense, inasmuch as it cannot but involve the mode historically adopted by man in considering himself and his relationship with nature. As we have seen, it represents an escape from the limits and abstractions with which man has traditionally formulated the problems of knowing himself as both subject and object.

To put it another way, man must overcome, once and for all, every "humanistic" prejudice as he reflects on his own history and his relationship with nature; and this is true both for the problems of knowledge and for those of praxis, scientific or otherwise. He must finally succeed in historicizing himself and his own history, in the process of evolution of which he is a part; and return to a path from which he has too long diverged. That is to say, he must begin again from a broader and more comprehensive notion of himself and his relation with nature. To put it in a nutshell, he must become inward with a level in that relation which, by the very fact of its own inner-ness, is deeper and more conscious.

Biological and historical evolution has always expressed a tension, mostly unconscious; when this leaves its static equilibrium and moves towards the level of consciousness, it always produces important cultural changes; this is historical emergency. At the point we have now reached, the emergency demands a new understanding—an explicit recognition of the inner workings of comprehension and performing, of self-making, that belong to evolutionary forms, as they *also* belong to man and his history.

All this corresponds to the necessity manifested in current scientific knowledge which is tending to shift the point of view further inside the cognitive process. This shift aims, as it must, to be more and more able, through scientific knowledge, to recognize the patterns of its own understanding and self-making.

There can be no doubt but that this reinforcement of knowledge, that is predicated in the historical conversion we are concerned with, will enable us to confront the new problems of equilibrium and of finding fresh perspectives, both in research and its implementation, which will only be possible, however, in the light of a re-established, systematic nexus of nature and knowledge.

Liberation from the "humanistic" prejudice and the abstract vision of human dominion that results from it, will free mankind from a distorted and enfeebled conception of a history subject to him alone. He will be enabled to overcome the conception of his will and knowledge merely as an isolated point from which to know. He will retrace his will and knowledge to the multiple evolutionary forms of his self-making and self-comprehending. In this way mankind can grow anew, not according to the terms of a dominion over nature that we have seen to be abstract and ultimately bankrupt, but rather as belonging in nature, in what it knows and what it does.

NOTES TO CHAPTER 3

1. Badaloni, (1970), p. 90 (my translation).
2. *Ibid.*, pp. 83 and 85.
3. In the *Dialectic of Nature* Engels argued that there must be a clearer distinction between selection by pressure of plant and animal overpopulation and selection by superior ability to adapt and alteration of surroundings. His declared intention was to tone down the "Malthusian" content in Darwin. What worried Engels was that an impoverished conception of the "struggle for existence" (impoverished, i.e., as against the complexity of historical development) might subsequently be transferred from the history of nature to the history of society. (Engels, 1935, Eng. ed. 1982, pp. 306–7). On the relations between Engels, Marx and Darwin, see, among much else, Gerratana (1972), chap. II, Di Siena (1972), Enzensberger (1973), Heyer (1982), Müller (1983), Vidoni (1985).
4. Engels 1935, English ed. 1982, p. 196.
5. *Ibid.*, pp. 196–7.

6. *Ibid.*, p. 198.
7. *Ibid.*, p. 236.
8. Badaloni (1976), p. 36 (my translation).
9. Engels, 1935, Eng. ed. 1982, p. 252. On Engels's criticism of mechanism see also Geymonat (1971), pp. 347 et seq.
10. Engels, 1935, English ed. 1982, p. 69.
11. Badaloni (1976), p. 35 (my translation).
12. *Ibid.*, p. 33.
13. Engels, 1935, English ed. 1982, pp. 69 *et seq.*
14. *Ibid.*, p. 212.
15. See Prestipino (1973), p. 186 and Vaiani (1971), pp. 80–83. On the formation of the young Lukács see also Rusconi (1970). As regards the more general problem of the dualism between social dialectic and dialectic of nature, some, such as Glucksmann (1971), retrace it to a whole tradition of leftwing idealism; others prefer to speak of a "differentiation" that can be found in Marx himself (Lombardo Radice, 1978, p. 17). I agree with Cantimori that a dualism of this kind is actually alien to the thought of both Marx and Engels.
16. Lukács (1969), p. 156 and *passim;* Abendroth-Holz-Kofler (1967).
17. Prestipino (1973), p. 184 and Abendroth-Holz-Kofler (1967). Compare also Apostel (1968), p. 38.
18. Gramsci (critical ed. 1975), p. 1345 (my trans.).
19. Prestipino (1973), p. 167. On the man-nature relation in Gramsci see, among others, Sabeti (1958) and Nardone (1971).
20. Gramsci (critical ed. 1975), p. 1874 (my trans.).
21. *Ibid.*, p. 1442. Typical and unmistakable is the example Gramsci takes concerning electricity: "Electricity is historically active, but not as a mere natural force (like the discharge of lightning that starts fires, for example), rather as an element of production dominated by man and incorporated in the totality of material forces of production; an object of private ownership. As an abstract force of nature, electricity existed prior to its reduction to force of production, but it did not operate in history and remained a matter for hypothesis in the history of nature (and prior to that, it was a historical 'non-entity', since no one bothered with it or even knew of its existence)". *Ibid.*, pp. 1443–1444.
22. In this connexion see the remarks of Badaloni (1962), p. 240, and (1966), p. 327, and compare with those of Lukács in Abendroth-Holz-Kofler (1967).
23. Cf. Ceruti (1986), p. 68.
24. Le Moigne (1982), pp. 179–80 (my translation).
25. Von Glasersfeld (1984), p. 20.
26. Gargani (1986), p. 9.
27. As a matter of fact we can note here a long theoretical-scientific process, starting from the researches of Needham, Weiss and Bertalanffy; it subsequently acquires consistency in the work of Piaget and Foerster, and makes use of the "second cybernetics". But it is above all in these last twenty years that the results of research in various fields—from neurophysiology to theories of cognition, from evolutionary biology to systems theory—have combined to make up the complex web of interdisciplinary relationships that underpin the achievements of epistemology. I have already mentioned some of the most representative works in this area of research; others will be referred to below.
28. Von Glasersfeld (1984), p. 31.

29. Concerning the origins of the "second cybernetics" see the contributions by Yovitz-Jacobi-Goldstein (1962). On the importance assumed by "second order" concepts in the biological sciences and the sciences of cognition and systems (self-organization, self-replication, autopoiesis, etc.) see the work of Zeleny (1980) and (1981), Depuy-Dumouchel (1983), Morin (1980), Varela (1979), Maturana-Verela (1980).
30. The resulting change in "function" and meaning of these concepts is well reflected in studies like those of Prigogine-Stengers (1979), Morin (1977), Jantsch (1980), Livingstone (1984), Watzlawick (1981), Laszlo (1986).
31. Ceruti (1986), pp. 107–108.
32. These levels appear specific, and relatively autonomous would seem to be the nexus that characterizes them, precisely because these levels are *tendential*, i.e. constructed on the basis of their transformability.
33. On the meaning of the notion of singularity see Thom (1980), p. 84.
34. On this change see the remarks of Thom on mathematics and physics (Thom, 1980, pp. 86–88 and *passim*), the reasons given by Jacob for proposing the *bricolage* model (Jacob, 1977, pp. 1161–66, and 1983) and the observations on the history of scientific thought in Serres (1977).
35. The notion of paradigm, introduced here, obviously refers to Kuhn's theorizing (1962). For the various implications of this problem see also Lakatos-Musgrave (1970), especially the articles by Masterman and Feyerabend.
36. On the Galilean "break" see Feyerabend (1978), pp. 99 *et seq.*
37. Thom (1980), pp. 84–86.
38. Ginsburg (1979), p. 92 and *passim*.
39. On the dualism arising out of the Galilean conception of nature, its reflection in Cartesian dualism and its more general implications, a passage of Husserl deserves to be quoted: "In his view of the world from the perspective of geometry, the perspective of what appears to the senses and is mathematizable, Galileo *abstracts* from the subjects as persons leading a personal life; he abstracts from all that is in any way spiritual, from all cultural properties which are attached to things in human praxis. The result of this abstraction is the things purely as bodies; but these are taken as concrete real objects, the totality of which makes up a world which becomes the subject matter of research. One can truly say that the idea of nature as a really self-enclosed world of bodies first emerges with Galileo. A consequence of this, along with mathematization, which was too quickly taken for granted, is [the idea of] a self-enclosed natural causality in which every occurrence is determined unequivocally and in advance. Clearly the way is thus prepared for dualism, which appears immediately afterward in Descartes.
    In general we must realize that the conception of the new idea of "nature" as an encapsuled, really and theoretically self-enclosed world of bodies soon brings about a complete transformation of the idea of the world in general. The world splits, so to speak, into two worlds: nature and the psychic world, although the latter, because of the way in which it is related to nature does not achieve the status of an independent world" (Husserl, 1954, English ed. 1970, p. 60).
40. Engels, 1886, English trans. 1968, 4th ed., 1977, p. 599.
41. Badaloni (1970), p. 88.
42. Gramsci (critical ed. 1975), p. 1449.
43. Lombardo Radice (1978), pp. 18 *et seq.* (my trans.).
44. *Ibid.*, pp. 21–22.
45. Badaloni (1970), p. 91 (my trans.).

46. *Ibid.*, pp. 81–82.
47. This notion must be referred to the historical distinction of four major social formations as proposed by Habermas: the traditional formation is the second of these and corresponds to the long period of pre-capitalist society; the modern formation comprises capitalist societies, based at first on free exchange, subsequently on organized capitalism, plus societies with planned economies. (Habermas, 1973, English trans. 1975, pp. 17–24). For further references see Parsons (1966), Riedel (1969), Dobb (1946), Hobsbawm (1962).
48. Badaloni (1970), p. 81 (my trans.).
49. *Ibid.*
50. It is worth mentioning the great importance of the theme of man's reconciliation with himself and with nature in the work of Marx and Engels (from the *Manuscripts* of 1844 to *Capital*, from the *Grundrisse* to *Antidühring*). Among the many studies on this subject, see Timpanaro (1970), Fiorani (1971), O'Malley (1966), Krader (1968), Stedman Jones (1972).
51. Prestipino (1973), p. 156.

CHAPTER 4

Natural Sciences and Socio-Historical Sciences

In the foregoing chapters I have pointed out the reasons why a thorough-going criticism of the "humanistic" prejudice that has for so long characterized historical knowledge is both possible and necessary. This prejudice has survived various changes and upheavals, even radical ones, in the conception of history. Indeed, it would appear to be one of the most important and persistent features of Western-Christian culture: a culture that has isolated man, setting him apart from nature and against it. So that history has become, as it were, a theater of actions and ideas of an abstract and solipsistic humanity.

As a result, the conception of the individual and society has been persistently abstract and artificial, this in the search for abstract foundations (theological, ethical, political) for the actions and history of mankind. And this historical "reduction" has survived even the scientific revolution with its Galilean paradigm. The scientific revolution banished that conception of history to a different realm by itself and the relegation was felt to be perfectly right and proper.

The adoption of the Galilean paradigm involved a sharper distinction and separation between the sciences of nature and the sciences of man, granting to the former a strong scientific status and demoting the latter to a status of scientific weakness and thus pushing them onto the fringes to make way for the new parameters of scientific knowledge. The result was that history paid twice over for its "differentness", and in the last analysis its "humanistic" prejudice was confirmed.

In the two centuries that followed, with the functionality of the new sciences and the kind of rationality that invested them—as against the slow and complex formation of the capitalist system and the laborious but inexorable rise of the middle class and its culture—the "humanistic" prejudice did not disappear. In passing from a conception of "progressive appropriation of the world" to one of "progress that masters the world", it altered its ideological bias and became even stronger.[1]

Far from having any adverse effect, the naturalistic objectivism of the "new science" and the repositing of the subject-object dualism in knowledge, and in the knowledge of mankind itself, actually reproduced "humanistic" history in new terms.

Moreover, it was the natural sciences, directly invested with the new rationalism, that benefited from the escape from the metaphysical-transcendental bondage; whereas history emerged as scientifically delegitimized, according to the very parameters of the new scientific reasoning.

So the "humanistic" prejudice of history persisted; and by favoring its relegation to a realm other than that of strict science, this "reduction" ensured that the prejudice continued to make itself felt. Indeed, with a lot of adjusting and adapting, the humanistic prejudice even put up tough resistance to the second scientific revolution.

One may dissent from the most radical interpretations that view the birth of contemporary physics in the framework of a general crisis of classical rationality at the turn of the XIX century tracing certain of its premises further back in time; but there can be no doubt as to the particular kind of change that occurred in that period and the growing importance of the arguments about the plausible links between notions of probability, of time and of certain knowledge.

In a wider historical context, we cannot consider these problems separately from the more general debate on theory that went on in the United States and Europe in the last decades of the XIX century and the early years of the present one and openly shed doubt on the Newtonian conception of a time *verum et mathematicum*.

Bradley, in *Appearance and Reality* (1893), as in *The Principles of Logic* (1883), explicitly rejected the conception of a unitary and global time, replacing it with one of several parallel times that referred to a series of different events simultaneously occurring. Thus the present became only a mental line drawn through the flux of events in order to connect one event with another occurring subsequently.

James, in *The Principles of Psychology* (1890), preferred not to speak of a single reality, but rather of "sub-universes of reality" (finite regions of significance), each endowed with its own specific time regime: as many local times, multiple, not co-ordinable unless abstractly, in a global chronology.

And along with the units of time, the ordered, "intransitive" succession of past, present and future was also called into question. The present, no longer a rigid point of demarcation between past and future, could be travelled in either direction. Is not the future "perhaps contained in the past, and must not the past change in order for the future to change?"[2]

In Freudian psychoanalysis, the past alters as we work over our memories. In any case, the past is present only in so far as it awaits a reply (from the future), otherwise it detaches itself from us.[3]

The philosophy of Nietzsche endeavors to deprive the present of the separating function it continually establishes and to connect it with an aleatory movement of becoming in a non-linear configuration.[4]

What we have here is, substantially, a sort of "hybridization" of the three dimensions.

The classical concept of time is attacked from another point: its scansion into equal parts comes in for criticism, and not only at the psychological level. With the Bergsonian idea of "duration", time becomes internal rhythm, elastic, an élan or contraction of an effort, a different way of ordering one's life and experience.[5]

We must note, however, that these reflections, diverging or critical with respect to the unitary and absolute time that was relegated to classical rationality, were developed at very different levels; so that when it came to tackling problems that belonged more properly to physics—as in the debate between Bergson and Einstein at the Paris Philosophical Society in 1922—difficulties and misunderstandings bristled. Nonetheless, the summations of these "relativizations" and new points of view, their convergence towards a more general "break" in the conceptions of space and time of classical rationality, remain of great historical significance.

All the more so in as much as that "break" was generalized and applied a jolt to the whole conceptual horizon: at the start of the twentieth century nothing was any longer intact and the crisis, as we may call it, was reflected in every aspect of cultural life and the dialectic of ideas. Indeed, it is awesome to watch how the changes came helter-skelter in every area in the space of a very few years.

One cannot but remark the fact that *annus mirabilis* 1905, the year of the special theory of relativity, stands out only as one among the little group of even more "wondrous" years in which raged a crisis of culture, scientific and otherwise, without precedent in history.[6]

It is well known that Einstein's theories remained anchored to a substantial "timelessness" of physics: "for us, convinced physicists, the distinction between past, present and future is an illusion, although a persistent one."[7] Nor did Einstein abandon a conception of time that included reversibility as a basic ingredient. Nonetheless, it is from him, in·a sense in spite of him, that we first learn of man himself being immersed in a universe in evolution. A closer look shows us that the teaching is analogous to that of Darwin or man's place in biological evolution.[8]

Newtonian physics set itself the task of describing a nature that was universal and objective since no reference was made to its observer; a nature that was complete, since at bottom it escaped the domination of time. One of the fundamental ideas of Einstein's relativity was, on the contrary, that scientific description ought to be consistent with the definition of the means theoretically accessible to an observer.[9] The observer belongs to this world and his position and physical constraint cannot be left out of account.

Following Einstein, and in spite of him, Heisenberg "grounded quantum mechanics on the exclusion of what the quantum uncertainty principle defines as unobservable".[10]

In the last analysis the "demonstrations of impossibility" both in the case of relativity as in that of quantum mechanics meant that the idea that nature could be described from the "outside", and that the cognitive point of view could be that of an "abstract" spectator, was a thing of the past.

But now, with relativity teaching that the physicist could no longer "idealize" an observation from without the universe, quantum mechanics dealt a second crushing blow to that "idealized" knowledge. Not only did it place man within nature, it assigned him a condition of "heavy" being, composed of a macroscopic number of atoms.

"It is a physics that presupposes an observer situated within the observed world. Our dialogue with nature will be successful only if it is carried on from within nature"[11].

Thus man can at last recognize the dual nature of his role as actor-spectator.[12]

And so it becomes possible to reconnect what classical physics had separated and opposed: the disembodied observer (abstract subject) and the object described from a position of exteriority (objectivized object).[13]

All this has another important consequence: "There is no scientific activity that is not time-oriented. The preparation of an experiment calls for a distinction between 'before' and 'after'. It is only because we are aware of irreversibility that we can recognize reversible motion"[14].

And it is this very admission of an irreversible time, directed towards the future, that enables the bridging of the gap that the "intertemporal" time of physics had opened between the "two cultures".[15]

Our physical conception of time has been further enriched by the subsequent progress of physics, including the evolution of the latest theories of thermodynamics.

Nowadays physics can no longer be said to deny time or its direction. It recognizes "the irreversible time of evolutions towards equilibrium, the rhythmical time of structures whose pulsation feeds on the fluxes that permeate them, the bifurcating time of evolutions through instability and the amplification of fluctuations, and even time. . . . that manifests dynamic instability at microscopic level. Every complex being consists of a plurality of times, each of which is linked with the others with subtle and multiple articulations"[16].

To all of which we must add the results—of no small importance—achieved by contemporary biology as regards the determining role of organization phenomena arising from irreversibility.

At a more general level of the "physis", then, we can speak of structures formed by virtue of an evolution such that its activity is the product of its history.[17]

Thus, within the sciences of nature there rears itself once more the problem of the historicity of nature itself, of its capacity for development and innovation, the problem of a nature capable of producing men and human society. All of which means that man stands on the threshold of a new understanding of nature, such as enables him to recognize himself as the product of nature.

But, in practice, this path leads back to the query posed by Engels: is it possible to reconstruct history and the sciences of man on a materialist basis?

As it turned out, the progress of science, which Engels a century ago saw as leading in the direction of a positive reply to his question, was not sufficient to achieve that theoretical goal he had hoped for: the principles of science were not substantially modified and there was none of that comparing and contrasting the results of one science with those of another that we have seen to be necessary. How different is the present state of scientific development![18]

Thanks to the principal achievements in theory of the second scientific revolution, the conceptual innovations that research has arrived at, the latest developments in physics and the astounding results of contemporary biology, our vision is an altogether new one with respect to that of classical science: *"a view in which the activity of questioning nature is part of its intrinsic activity"*[19].

If this fundamental statement is now perfectly consistent and arguable—as indeed it is, in all its implications—this must mean that it has become effectively possible to re-establish what I have indicated in this book as the continuity of the man-nature relationship, conceived in the new terms of a nexus of nature and knowledge that is both basic and systematic.

The consequence is of no small moment. The history and sciences of man can no longer be categorically separated from the history and sciences of nature: this cannot happen at the level of a categorical distinction-opposition, which—in the conspectus to which classical science and reasoning had accustomed us—led us to a definition, as rigid as it was abstract, of the different levels of reality; nor can it happen in the sense of an even more artificial distinction between different levels, so that human history and sciences are relegated to an "other" plane, sharply and abstractly separated off from the plane where the history and sciences of nature belong.

A quite different distinction of levels of reality, of their "self-building" in evolution and of the autonomy that justifies their belonging to the area of scientific analysis, actually credits scientific research and the multiplicity (often interconnected) of its orientation with the possibility of reaching a unified perspective; and at an appropriate level of complexity, it enables us to tackle the problems of the theory of knowledge therein implied.

From this standpoint and in view of all the foregoing, it would seem possible to state—in general terms—that the finiteness or, if we prefer, the relativeness and particular nature of the cognitive situation in which man finds himself from time to time, or in which he is "historically" located, can no longer be considered as a source of error.[20] Indeed, it is the "singular" and relative situation itself that constitutes the starting point and the point of relation with every other possible term of reference in cognitive activity; and this relationship and its intrinsic character inscribe themselves within a substantial continuity that is guaranteed by the systematicity of the nature-knowledge nexus at all degrees of evolution.

But if the historicity (or relativity) of this cognitive situation can no longer be substantially discriminated, as regards its epistemological valence—whether it be applied in fields belonging to the socio-historical sciences or in those of the natural sciences—this does away, in effect, with every obstacle, every paradigmatic barrier that has served to separate the two fields. There is no longer any reason why the socio-historical sciences should labor under a weak scientific status or be prejudicially differentiated, to their disadvantage, from the natural sciences. There is no longer any serious scientific foundation for regarding history as a "different" sort of science—indeed, not really a science at all.

The continuity of the man-nature relationship has been re-established in the new terms demanded and suggested by the most important developments of contemporary science; at this level, and with respect to the implications inherent in the theory of knowledge, it marks a definitive overcoming of that relegation of the history and the sciences of man to an "other" plane. Historical knowledge can thus be liberated from the "reduction" that derived from the condition to which that knowledge had been demoted; it can cast off that weakness of scientific status to which it appeared irredeemably condemned by its paradigmatic separation from the natural sciences.

It remains to explain, then, how and to what extent the "humanistic" prejudice, or the difference of condition of the history and sciences of man, have remained unaffected by the developments of the second scientific revolution: for it is a fact that they have survived the break-up of the entire framework of certainties that belonged to classical science and classical rationality, as they have survived the gradual growth of an entirely new perspective—that which envisages a possible relinkage of human sciences and natural sciences.

I shall not neglect to give some account of the considerable differences in positions on this matter, distinguishing that which belongs to the genuinely philosophical argument from the connotations of historical research as such. But it must be said at once—if only by way of summary and general evaluation—that little has been made of the considerable opportunities to elaborate and renovate the methodology of historical research, offered by direct and explicit confrontation with the results and theoretical implications of scientific developments in the first decades of this century and later.

Another fact is highly significant. Even if we confine ourselves to the later nineteenth century and the first half of the twentieth, we find a wide-ranging debate between philosophers on the scientific character of historical research (or lack of that character). But with some notable exceptions—namely, historians who were also philosophers—the general run of historians has made no important contribution to this debate.

True enough, the reference to historical research in that argument was often instrumental, or underlay the more general analyses of the structure of reality and the problems of our knowing them: but this scarcely justified an attitude of caution or mistrust towards a theorization that might sometimes appear in the role of "usurper". Still, this attitude was

only too consistent with historians' approach to the problems of historical knowledge and its "scientificity"—a stance that for long remained
defensive.

For decades historians preferred not to expose themselves in discus-
sions over theory or open arguments about methodology; and they even
veiled the most traditional methods of interpretation they continued to
employ, refraining not only from expounding them but even from defend-
ing them.

In effect, they fell back on the specificity of their discipline: this retreat took the form of an attempt to trim their methodology by making it
more precise and more refined.

This attitude, so widespread and so durable, gives the key as to how
and why history has remained on an altogether different level from that
of the other sciences. The inadequacies and substantial weakness of that
position have loomed large even when set beside other social sciences
that have at some point gone in search of a sounder theoretical framework. Thus we see how the "humanistic" prejudice still holds firm in
the practice of historical research.

This, then, is how things stand and threaten to continue. The important contributions to methodology of the second scientific revolution
have not been faced up to; nor should we be surprised when we find
historians thus adopting a passive standpoint as philosophers and philosophers of science argue over the requisites of scientificity in historical
knowledge.

Face to face with the positions that emerged from that long debate—
positions as different as could be, even in direct contrast to one another,
the historians' reactions showed no change: a solution was to be sought
within their particular problems, not outside them.

At the same time it must be said that the debate over the scientificity
of historical knowledge went forward in such a way, in its main lines and
most significant positions, that in the end it seemed to encourage the
defensive standpoint taken by historians and their contention that history
is different. And in the last analysis this is all too understandable in view
of the fact that the traditional separation between human sciences and
natural sciences rested on basic conceptions that had long been consolidated and were not easily overcome. To supersede them required a
profound and radical re-appraisal of categorical distinctions and methodological differentiations. Direct methodological confrontation could only
come from a courageous and unprejudiced re-consideration of those
achievements of the second scientific revolution and its broadest perspectives which make for convergence with the sciences of man. But we

have seen how in the area of human sciences and their possible re-foundation, the opportunity for a revolution has substantially been lost or, at most, has been a long time taking shape—this in spite of some of the essential preconditions already being there.

Thus it is easy to understand how prior to that revolution, in the period when positivisitic theories ruled unopposed, a dual attitude towards historical knowledge prevailed, differentiated but leading to the same result.

On the one side stood the positivists who insisted on an ideal of scientific knowledge modelled in nomologic and deductivist terms; they demanded, since they maintained it was possible, that historical research be fitted to that model, taking for granted the structural unity referred to in their scheme.

Obviously it was no mere accident that a position of this kind—of which Buckle was one of the leading exponents[21]—appealed only to a minority. And although it was taken up and much elaborated by Hempel[22], it produced no conclusive results, since historical research continued to labor under a weak scientific status: its backwardness could not suddenly be compensated in a single leap, by a sort of decree of logic, nor by purely and simply conforming to a "different" model that remained just that.

Other positivists adopted a less unilinear, more diffuse position. With the same view of scientific knowledge before them, they allotted historical research a particular place in the system of the sciences, acknowledging the special difficulties and the reasons for which the criteria of scientificity of historical knowledge could not be tailored exactly to fit that system. But by so doing, what though from their positivist point of view, they ended by endorsing the "diversity" of historical knowledge (and thus its inferior status) which could continue to be upheld.[23]

The historians' reactions to both of these positions—a reaction even somewhat encouraged by the second—could not but stem from the wholly "internal" logic of their typical defensive attitude.

It led them to seek to perfect the criteria of consistency and rigor in their methods of proceeding: the critical evaluation of sources, the use of documentary data, the care given to systematizing these, the caution and scrupulosity in interpretation. In all this we may detect a return to, or a residue of, the traditions of scholarly history, whose influence is still by no means dead. There was also a franker attempt to bring methodology up to date; but this effort was applied rather to particular procedures that had already been consolidated than to any courageous re-appraisal of the condition and the scientific status of history, and it remained more of a mimesis, an attempt to fit history to the dominant scientific model, instead of a real change of directions and methods.

Historians let themselves be influenced by positivistic ideas in quite a special way, attempting to render their standard criteria scientifically more valid while sticking to their basic conceptions; these latter remained very far not only from the ideas of the positivists but also, more generally, from the foundations of natural science.[24]

This kind of reaction, defensive and almost mimetic, with its attempts at adjustments, but fundamentally consistent with the peculiarity of the discipline in question, is important for two reasons. Firstly, it erected a sort of protective wall—that could be crossed, but never entirely vanished—an emplacement, as it were, that historians continued to hold even after positivist ideas had ceased to hold sway. The position was maintained even when positivism exerted a belated, indirect influence through logical neo-positivism and, more generally, through its role in epistemological research. The second reason why that reaction and the kind of adjustment that characterized it worked so successfully, was that, simultaneously, the defensive reaction gradually found new support and was able to maintain its position of discipline in its own right, thanks to the support it also received from all antipositivist and antinaturalist theorization.

According to the latter it was quite out of the question to speak—as had the positivists—of a subordinate character of historical knowledge with respect to the system of the sciences; if anything, they emphasized the absolutely special nature of the object of that knowledge. In this way the diversity of the historical process, as structure and methodology, was reaffirmed and raised again.

Nor, indeed, had the historians ever really departed from that conception of diversity and it was only underlined all the more by the positions taken by neo-Kantians, by the theories of Windelband on historical research as description of individual facts, by Rickert's reminder of the special nature of historical research as necessarily referring to values.[25]

To say nothing here of neo-idealist conceptions. Running to the opposite extreme in the reaction against positivism, these actually raised the "humanistic" prejudice to a pride of place, with man supreme, in his ideas, his actions, in all historical becoming.[26] As we know, this supremacy was to invest the "special" nature of historical and philosophical knowledge, with the other kinds of knowledge demoted to inferior positions. The "different" nature of historical knowledge, as we have described it here, thus received its greatest emphasis and was sanctioned by being caught up into an all-embracing system of philosophy.

And what is the upshot of this long philosophical dispute over the scientificity of historical knowledge? It can be detected largely in the latest theorizations, or at least those put forward after neopositivism had begun to relax its domination; and it can be summed up as a trend towards a solution of compromise.

The tendency here is to grant historical knowledge a place all its own within the system of the sciences. However, this collocation does not come about as a result of history's simply being assimilated to the connotates of natural sciences; this is not scientificity in a strict sense: rather, it is derived from a more general concept of science and from a more specific reference to the social sciences. The aim is to formulate criteria of scientificity that shall be equally valid but better suited to their object and to the specific requirements of historical research.[28]

And so, travelling by this path, the diversity of historical knowledge, although scientifically reappraised and reworked, once again finds confirmation—and, if necessary, with even more evidence in its favor. Indeed, the dominant logic of these solutions is the same old one that rules out any unified perspective. Obviously not in the sense of a presupposed model of positivist or other mold—there is no further demand for that; but rather in the sense of a close methodological comparison based on actual research results, in both areas, and thus effectively capable of innovation and re-linking in the relationship within the scientific system.

If anything, the newest element to emerge from these recent approaches is another one: the reference to social sciences, and thus the reconsideration of the problem of historical knowledge in a broader conspectus—what though this also concurs in a basic distinction between the sciences of man and the sciences of nature.

The novelty also consists in its contribution to the renewal of historical research; but that is not really saying much.

From this point of view one of the most significant theorizations was offered by Weber in 1904 in his essay on "Objectivity". Weber's positions deserve attention for various reasons. First of all, he approached the subject from a direction other than the philosophical one; indeed, he was in reaction against those philosophical conceptions that tended to invest history with a general sense. The reaction occurred in a very precise context (also as regards time) and I have already mentioned how it broke with the scientific and philosophical certainties of the period immediately preceding it. Weber, however, stood apart from other thinkers of his time who more radically denied that history could have any sense. Not only was Weber's position a more constructive one, though in a different way, but those attacks on the sense of history were still made in a philosophical key, what though the philosophy ran in a different direction.

Weber's point of view related, rather, to an entire evolution of the social sciences which, from the second half of the nineteenth century onwards, had been in search of a theoretical justification and a conceptual systematization.

The other reason for being interested in Weber's conceptions is that they have found favor not only with social scientists but with quite a few historians as well. Among other things, they met the demand, quite widespread at the time, for a critique (revisionist or thoroughgoing) of Marxism; this critique, however, shied away from other conceptions that were, rightly or wrongly, held to be equally "all-embracing".

In the essay mentioned above and which we shall examine here, Weber argues that historical knowledge, like other kinds of knowledge, could not exclude reference to "values".[29]

The only scientific "objectivity" available to the researcher lay in his capacity to take account of a plurality of systems of reference endowed with values; the affirmation of these values being dictated by the tendency of social conflicts.

From these premises Weber then attempted to extract a specific methodology of evaluation, on the basis of which single affirmations were possible. The possibility of a scientific justification, in the strict sense, of these evaluations was still ruled out[30].

In sum, Weber viewed the socio-historical sciences as distinguished from the natural sciences by the reference to values, present in the former as they were absent in the latter. And it was this very reference to values that determined the significance of a historical object and thus its individuality. Since the method of the socio-historical sciences remained anchored to their orientation towards individuality, what was the distinction between them and the natural sciences, at structural and methodological levels, if not absolutely clear-cut?[31]

And thus this point of view further confirmed the distinction-separation between the two types or levels of science.

At the time and afterwards these positions enjoyed the reputation of a certain modernity and critical awareness and, as we have seen, they ended by exerting some influence on various historians. Their success was, to be sure, partly due to the fact of their being able to coexist and combine with other formulations and orientations; but there were other reasons.

In the formulations of historians both long past and very recent, we find again and again an appeal to the plurality of systems of reference with respect to which "historical pronouncements are relative"[32], with the inevitable corollary that these systems of reference—unlike those of the natural sciences—are endowed with values; what is actually happening here, though perhaps in more up to date terms, is an ever sharper separation of history and the sciences of man onto a "different" plane.

The main feature of that "diversity" stands out, the relativity of pronouncements is reaffirmed and the reference to systems founded on values, which must always be different, is set against the "objective" reference of a science that is able to dispense altogether with value relations.

So this path too ultimately leads to distinction-separation. The comparison between the methodologies of the two types of science once again misses its target of a truly innovatory solution that might lead towards a unified perspective.

The various positions we have hitherto examined have revolved around the problem of the relationship between history and the human sciences, on the one hand, and the natural sciences on the other. It must be said that the formulation that gave most promise of developing in a way consistent with a unified perspective was, without doubt, that of dialectical materialism.

We have seen how the achievements of the second scientific revolution and certainly theoretically important trends in its subsequent developments, both in physics and in contemporary biology, have once again raised the question of the "historical" character of nature within the ambit of the selfsame sciences of nature. The problem had already been posed by Marx and, especially, by Engels in the *Dialectic of Nature*.

All the recent developments in physics and biology have produced results whose theoretical implications tend to reconsider the man-nature relationship in terms such that—on the more general level of the theory of knowledge—"understanding nature meant understanding it as being capable of producing man and his societies"[33].

Which sends us back to the formulations of dialectical materialism and the demand, advanced by Engels, that history and the sciences of man be reconnected with their materialist basis in order to reconstruct them on that basis. Today we find ourselves facing that problem again; but this time we can address it in much more effective terms than were possible with the state of scientific knowledge in Engels's time, thus improving on the theoretical formulations he envisaged.

Science has developed in such a way since then as to validate the demand put forward by materialism. This can obviously be conceived in different ways. Nevertheless, as a theoretical exigency, it can be recognized as that most closely corresponding to the change in conceptual position demanded by the scientific results we have mentioned.

Clearly the need is for a modern materialism, envisaged in different terms, able to enter into a dimension internal to the sciences and become a moment in their constitution. On the other hand it should not be forgotten that—from our point of view—the most modern, most significant aspect of Engels's formulation consisted in the direct reference he made to the results and findings of the sciences. It was only by virtue of these that the problem could be propounded and, thereafter, its solution be indicated most completely and exactly. In this connexion he states unambiguously that science, just because it is science, will ever more imperiously take the place of philosophy.[34]

Between the potentialities of this first approach, this hopeful propounding of the problem, and its scientific relevance, lay a gap that for the time being was not closed and could not be. Marxist scholars became entangled in the barbed wire of the problem. We can gauge the price they paid in the course of a particularly tormented debate, which saw a regression to the most rigid dogma, the most schematic traditionalism; which saw some dissociating themselves from the Engelsian position or adopting the most widely contrasting positions.

But the upshot of all this was that the possibilities of a new elaboration or a critical reappraisal faded—either because of the stand taken on dogma, or as a result of reaction against dogma, or for other reasons.

Turning now to Gramsci and Lukács, we note the highly problematic, complex nature of the positions of the former in the various passages in his work devoted to the man-nature relationship, and the tortuous route taken by the latter, in a part of his work that carries no small importance; both are eloquent examples of the real difficulties that arise in this connexion, and show a certain critical detachment with respect to the early formulations, a detachment that was to characterize a whole subsequent generation of Marxist thought. To this we must add the weight of historical factors connected with the Soviet experience, and thus, with the development of Marxist thought in the context of that experience. The dogmatic hardening that was a feature of the Stalinist period, and the schematic interpretation of dialectical materialism that resulted, could not but create profound divisions in the ranks of Marxist scholars.[35]

Nowadays it is possible to view these things from a critical distance, and we can easily see how neither the first kind of position nor the second could enable and adequate re-appraisal or reformation of the problem. The schematic impoverishment of the first and the critical diffidence or dissociation in the second led to a like result: a neglect of undervaluation of the most original, most potentially fertile elements in the early formulations of Marx and, above all, Engels.

Viewed in light of those possibilities, this later work can be seen to have come no nearer to its aim: an aim that was not only feasible but indeed stood as a premise to that work, namely a real elaboration of the continuity of the man-nature relationship and the reassociation of the two areas of science.

The most valid, most relevant aspect of Engels's teaching involves the direct reference to and comparison with results of scientific research; armed with this, we can go beyond the uncertainties and the theoretical shortcomings that also manifested themselves in the debate that followed.

Now that the most recent developments themselves in the sciences demand that the problem be repropounded, and make this a possibility, the conditions also exist for us to override the traditional separations of human sciences and natural sciences into different levels and different status. The conditions exist for regaining the objective of reattaching the socio-historical sciences to their materialist basis, in order to reconstruct this basis in new terms.

And this may lead us to reinterpret the demand for materialism on the basis of the results and findings of science; with the consequence that the results are given new orientation, being shifted within a dimension internal to science. In this way, not only is all dogmatic deduction excluded, but also the schematic superimposing of theory on real cognitive processes.

The new materialism I refer to is grounded in the systematic nature-knowledge nexus throughout the evolutionary process. And it is the paradigm of form that enables this nexus to be represented at each stage and level of evolution. On this capacity for representation and correspondence ultimately depends its verifiability in the context of research.

The conceptual horizon against which the unified perspective now emerges can be seen to be that of a general process of experimentation, physical, biological, cognitive, characterizing all natural and human evolution: more precisely, an evolution that itself advances by continual approximations, through a great many special situations, represented by form as active synthesis occuring at each level and each step of evolution.

The process fully confirms the specificity and historical determination that can be adduced in defense of the non-schematic, non-deductivist character of the paradigm proposed—a paradigm so much less imposed from above. And so, even in the new perspective, it can be stated that "the quality of process of the world and its materialness can be verified *inside* scientific research."[36]

Thus we can once again posit and pursue a unified perspective that finds in form a new paradigm able to interpret the developments of science that press towards that perspective; this points the way to findings of great significance for epistemology. It also puts an end to the condition of separateness and "difference" that has bedevilled history and the sciences of man; the condition to which they had been relegated to cultivate, for themselves, the basic "humanistic" prejudice.

History and the sciences of man were allotted a role within the system of relationships in science, and delineated this role as resolving itself into a position of substantial subordinacy, of logical and methodological minority. And the conceptions that, from time to time, supported and justified it, were of necessity misleading, erring, afflicted with a constitutive partiality.

This burden has made itself felt every time anyone has tried to justify, or find reasons and criteria for, a special scientificity that might correct or redress the weakness of the scientific status of history, in particular, or of the human sciences, in general, as compared to the "strong" sciences.

We have seen how this lop-sidedness actually came about, how it persisted on both sides and how, only later, the newest developments in natural sciences have led to results and theoretical inferences that have enabled the problem of a re-linkage to be posited once more. In any case, all attempts at readjusting and reinforcing are vain if made only on one side; apart from being vitiated from the start by the particularity of aim proposed—it was taken for granted that the object of historical research had a special nature—they could only be, at best, limited "in terms", partial, inadequate to the overall problem. By the same token, all efforts to make history conform to the dominant models and typologies in the other area, that of the "strong" sciences—for reasons opposite but symmetrical—could only end in an abstract attempt, as ineffective as it was mistaken.

What to do in the face of such marked disparities of methodology, such clear-cut differentiations, and even contrasts, of typology and procedure, with all the difficulties and delays they put in the way, the partialities they involved—especially in the case of a "weak" science like history, but also for the totality of relations between the sciences? What else, but to see in the end that these problems could be dealt with and solved in only one way, namely by the genuine re-linkage of human sciences and natural sciences; that is to say, in a unified perspective.

But alas, as we have seen, it was (and is) that selfsame historical knowledge, more than any other, that holds itself apart from any relationship or comparison of methodology that might, even partly, lead along that road.

Whereas on the other side things took a different turn, at least with the implications of the second scientific revolution and the surprising developments of natural sciences. Methodology was transformed, conceptual viewpoints profoundly changed; there was a demand for new theory—all this pointing towards a convergence that took on more and more the aspect of a unified perspective. Some of the most significant theoretical implications ended by directly affecting the area of the human sciences; but in the long term it was inevitable that these latter, especially history, stranded for so long in a condition of diversity, should break loose in their logic and methodology.

The comparison and confrontation with what is being worked out on the scientific research front has not come about. The trend has been towards overcoming the traditional divisions between different levels of reality, but historical research appears to have taken little notice of this. That is why it has not been able to reappraise its conditions and status, its placement within the scientific system. And this has led to an increasing marginality in the reorganization of the relations of that system that have been proceeding in the meantime.

In situations of this kind, of an essentially critical (crisis-oriented) nature, the ever-present danger is that of being demoted to a subordinate place. This means paying a very high price at the logical as well as methodological level. Indeed, the consequence is subjection to categories and parameters not only developed in other areas, but even already dispensed with in those areas as part of a general renewal, an overall advance in science. Not to have taken part in this means exclusion from its rewards and prospects.

In this connexion, the most macroscopic, most paradigmatic example we can adduce is that of the concept of cause and its inconsistent use in historical research.

The debate over criteria of scientificity in historical knowledge has been long and many-sided. Often, and from differing standpoints, it has been asserted that causal explanation is both possible and useful in historical research. The intention here was *de rigeur* to demonstrate and/or reinforce history's scientific character. Some positions have been most rigidly prescriptive, others more conscientiously bent on stressing the special nature of the object of historical research and its methods; but both were to find it equally hard to posit causal explanation. Even the most careful and critically circumspect attempts in this direction have, as a rule, ultimately recoiled from specifying how and to what extent causal explanations of history can actually be given. In almost all cases the result has been to advance a much reduced or trimmed version of this type of explanation and the terms on which it can adopted in historical research.

To begin with, Weber, as we know, felt that the socio-historical sciences should offer explanation of a causal type as did the other sciences; but his very formulation of the problem generated difficulties of no small order. For if these sciences must be distinguished from natural sciences by virtue of their necessary reference to value, the causal explanation of their object—the historical object, in our case—could only be a partial one. And so, from the classical model of causal explanation (capable of identifying necessary relationships) one had to fall back on a model of "conditional" explanation. Which was to say that in historical research one must make do with identifying a series of conditions, in the presence of which, but alongside others, a certain event could take place.[37]

Hempel, on his side, argued that historical research should follow a nomological-deductive model, maintaining rightly or wrongly that this is the model for any scientific procedure. But he had to admit that in the case of history no "finished" explanation was possible; the best we could hope for was a "sketch" for an explanation, which then required rounding out with further empirical data.[38]

More recently, and in a climate less influenced by neopositivist ideas, various philosophers of science have returned to the problem only to find themselves grappling with difficulties that ultimately turn out to be the same as above.

Von Wright, for example, had to admit that in history causal explanation gave rise to relationships "less close" than those occurring in the area of natural sciences. History cannot avail itself of universal laws; it employs, instead, a series of special statements which may be taken as premisses. The explicative typology deriving from these statements answers the question: "how was it possible for a particular action to take place?"[39]

The conclusion does not differ greatly from Fain's example of the watchmaker: the historian, like the watchmaker, is able to describe a certain functioning, but not the laws that lie behind it and explain it.[40] But even without taking up sharply antipositivist positions like Fain's, authors not averse, in principle, to a nomologic-deductive explanation— from Gardiner to Atkinson—end by replacing the general laws that hold in other scientific fields with special laws for historical research. And these special laws take on the much more limited connotation of "resumptive generalizations".[41] Thus if causal explanation is applied to historical research, it can only produce a very weak or limited relation (which can, at best, be configured in a sequence of the type: "a" therefore "b"; or of the type "action *a* took place because it was preceded by action *b*").[42]

According to Martin, explicative typology in history demands the adoption of certain rules, but these are especially conventional rules.[43]

In these different cases it would seem fairly clear that the sequence of events, as configured in the most typical procedures of historical analysis, can at most constitute a premiss for subsequent events. It is inevitably excluded from a closer causal relationship, and ends by appearing as a relationship weaker than those guaranteed by the models of explanation adopted in other scientific typologies.

This kind of conclusion cropped up quite frequently in the course of theorizing and found a place in the formulations and modes of procedure of historians. In general, those formulations stand very far apart from nomologic and deductive models of explanation[44] and are hard to reconcile with them; and not only that—very often they were, and remained characterized by the most traditional methods of proceeding, by a taste for highlighting the singularity of events and by a limited reconstruction of the sequence of those events in purely narrative terms.[45]

But—and here is the first striking incongruity—this state of affairs did not prevent (as it does not even today) historians from resorting habitually and shamelessly to the concept of cause. The use made of this is as inconsistent and uncontrolled as it is customary.

Historians tend to be uninterested or indifferent toward the study and elaboration of methodologies, and in a more general way towards the questions of theory that arise out of research. Nor does it seem to bother them when they adopt logical categories and conceptual devices that are alien or extrinsic to their own formulations and procedures.

So numerous are the examples one could give of the improper use of the concept of cause, that it might be more worthwhile to construct a table showing the recurrence of the word "cause" in a host of history books. In this way one could see how the concept is often left to oscillate between its most pregnant logical meaning and its degraded meaning in everyday language.[46] This is yet another demonstration of a more general fault that can be laid at the door of the historians: namely, that they have hardly bothered to work out the theoretical bases of the language they use.

But the example of the concept of cause now enables one to work out another, even more serious contradiction; one that throws more direct light on the logical and methodological backwardness under which historical research labors as compared to the progress of other sciences.

On a par with other concepts in classical reasoning, for a long time the concept of cause found a precise correspondence and foundation in the representation of scientific phenomena. The relation of equivalence between cause and effect, so that the effect can be retraced to the cause, corresponds to the character of reversibility of phenomena as depicted in classical physics.

In the mathematical description of motion, by Galileo and Huyghens, the relation of equivalence between cause and effect corresponds to the reversibility of dynamic trajectory: the example commonly given is that of a perfectly elastic ball that rebounds and returns to its point of departure.

Quantum physics introduced far-reaching conceptual innovations. The inadequacy of classical dynamics required that new concepts be brought in: an essential one is that of the operator. Generally speaking, we may say that this concept found its effective correspondence when the notion of dynamic trajectory was set aside, and with it the deterministic description implied by the trajectory.

The introduction of operators profoundly alters our basic concepts, leading to a type of uncertainty alien to classical thought. If the coordinate and momentum operators be ascertained, quantum mechanics does not allow a state in which coordinate q and momentum p both have a definite value. This situation is expressed by Heisenberg's famous uncertainty relations: if the coordinate is precisely determined, the momentum can assume any positive or negative value. Which means that in an instant the position of the object will become *arbitrarily* distant. "Heisenberg's uncertainty relation necessarily leads to a revision of the concept of causality".[47]

But leaving aside this highly significant case and speaking more generally: the fact is that "quantum mechanics deals with only half of the variables of classical mechanics. As a result, classical determinism becomes inapplicable".[48]

Another most important point is that the quantum phenomenon presupposes irreversibility. As Bohr and Rosenfeld often emphasized, when dealing with quantum phenomena "every measurement contains an element of irreversibility".[49] In the analysis of such phenomena the data cannot be obtained by means of idealization in an instantaneous space-time reference. If an electron leaves its stationary state and jumps from one orbit to another, only after this process, which is to say after an irreversible transformation, can we know the values of the various energy levels of the system. Otherwise, if the electron were to remain indefinitely within its orbit, we should not be able to describe it.

Thus it comes about in quantum mechanics that even when—as in the case of the Schrödinger equation—a reversible and deterministic evolution is expressed like the canonical equation of classical physics, that equation can be verified only by irreversible measurements that itself "is, by definition, unable to describe".[50] In this way reversible and irreversible evolution are inexorably intertwined, but it would still be equally impossible to expect to correlate all physical evolutions with a deterministic, reversible transformation.

Now we come to the supplementary problem posed in quantum mechanics, i.e. the coexistence of reversibility and irreversibility. This is indicative of the fact that the classical view, leading to a description of "the dynamic world as self-contained, is impossible to the microscopic level".[51].

In classical as in quantum mechanics, "irreversibility enters when the ideal object corresponding to maximum knowledge has to be replaced by less idealized concepts" such as can be described by statistical sets[52].

Hence the more general need, as expressed in microphysics, to adopt a probabilistic logic.

This overcoming of the concept of cause and adoption of a probabilistic logic had interest for the socio-historical sciences, too. It offered them the possibility of re-examining, in a conceptual context other than the classical one, the contradictions that arose when deterministic models of explanation were employed. This was no simple matter of swapping one model for another; rather it involved facing up to the overall significance of the complete turnabout in the progress of science, a significance that reached far beyond the mere field of immediate application.

Thus the probability model became much more than a model of explanation pure and simple like any other. It involved a whole shift in the conceptual perspective; and this was certainly not destined to end with physics, but would extend its influence throughout the subsequent development of the whole area of science.

Bohr himself was at once aware of the more general implications in that change of perspective as reflected in his principle of complementarity.

The result of measurement—as conceived in the new terms—does not give us access to a given reality. If the quantic number measured characterizes the system in the state in which we have *chosen* to describe it, this generates a *multiplicity of points of view,* and shows how far we have departed from classical objectivity; or, shall we say? it marks our refusal to give a complete description of the system as it is, independently of how it is observed. Hence the principle of complementarity (which can be considered as an extension of Heisenberg's uncertainty relations). The various ways of speaking about the system and the various points of view complement one another. They refer to the same reality, even though they cannot be subsumed into a single description.

And Bohr was right when throughout his life he insisted that the principle of complementarity had a more general significance for other fields of knowledge and could be translated into them.

In the final analysis this was the wider meaning of the turning point marked by quantum physics, underpinned by the move from deterministic logic to a probabilistic logic; but the comparison of methodologies that thus became possible, also in the ambit of the socio-historical sciences, could not confine itself merely to imitation or adjustment.

Here a reverse trend began, and with subsequent developments in physics and the relations between physics and the other natural sciences this inversion became more and more evident, until it resulted in a new relationship with the human sciences as well. Newtonian science had inflicted defeat on all the other scientific typologies that had shown themselves weaker and less corresponsive to the parameters imposed by that same Newtonian science; but now it began to appear that the defeat was not a permanent one. The 'strong' sciences and the 'weak' sciences had been sent along different, ever more diverging paths in a sort of scissors movement, with history representing the furthest relegation of the weak sciences; but now it began to appear that the gap could close.

And the confrontation of logic and methodology between the two scientific camps offered possibilities and values quite different from the terms in which that confrontation had previously been understood. This time it could not but be bound up with a new orientation of scientific development. Thus it was to take on the meaning of a convergence towards a new perspective in which the relations between the different scientific typologies were reappraised and redefined.

So that it was now not only possible but imperative for the socio-historical sciences, too, to re-examine their condition and scientific status. They could also reverse trend, emerge from the marginal, subordinate condition to which they had been relegated just by virtue of being different from the other sciences.

The logical and methodological *plateau* quantum physics had reached, constituted on a probabilistic basis, represented a striking innovation in terms of the grammar of scientific knowledge. The elements of the cognitive process now stood in different relation to one another: differently related were subject and object; different was the very sense of the predicate. The new logic underpinned, and was interpreted by, an entirely new formulation of the problem of knowledge. And this novel articulation of the grammar of knowledge offered values and meanings that were cogent also from the standpoint of the socio-historical sciences.

So it was not simply a question of confrontation, point by point, with a mere model of explanation. A confrontation of this kind could have been conceived, once again, in terms of utilization and adaptation pure and simple, even though this time round the model was innovative and otherwise probatory, as compared to the unsuitable, inflexible model of causal explanation. This might have brought some advantages in the way of a conceptual apparatus that was certainly more apt to the task, but it would not, by itself, have been sufficient to help the socio-historical sciences out of the condition of subordinacy to which the entire previous confrontation had relegated them.

But the most authentic significance of an open confrontation in the light of the turnabout we have described above was that it offered the chance for the socio-historical sciences to close the gap, to converge towards a perspective of change in the relations between scientific typologies.

But, as I mentioned above, the path that led to overturning the defeats inflicted by Newtonian science was destined to grow broader, to take other turns and to reach into new territories.

The latest developments in physics, especially in the thermodynamics of irreversible processes, the new contributions from physical chemistry, the progress of contemporary biology and, above all, the ever closer, ever more fruitful relations being established between one field of science and another, these achievements represent the outlines for a new and ampler prospect that is just as important as they are.

NOTES TO CHAPTER 4

1. On this subject see the remarks of Horkheimer (1930) and Horkheimer-Adorno (1947); cf. also Kittsteiner (1985).
2. Butler (1872).
3. Freud (1916–17).
4. As is well known, this concept of time recurs at various points in Nietzsche's work. But see especially the second of the *Thoughts out of Season*, where the problem of time is related to the meaning of history (Nietzsche, 1874).
5. Bergson (1922).

6. Russell's antinomy belongs to 1903 and in the history of mathematics marks one of the most serious crises in scientific thought. The rise of non-Euclidean geometries seems to have been accompanied by a general questioning of the more usual dimensions of geometry. Even in painting we find space disintegrating and reconstructed in a novel way: Picasso's *Les demoiselles d'Avignon* belongs to 1907 and heralds the beginning of cubism. And there are other artistic innovations of deep significance in those years, especially in literature. They show that in the midst of a general crisis the "subject" was up for reappraisal. Indeed, in psychology depth analysis seems to have brought about a kind of "Copernican Revolution" in the universe of the human subject: the *Interpretation of Dreams* and the *Psychopathology of Everyday Life* belong to 1900 and 1901 respectively. Lastly, we seem to have reached the end of "preestablished harmonies"; and not only those of the Newtonian universe governed by eternal unchanging laws. Other harmonies (in the literal sense, too) underwent revolution: for instance, atonal music makes its first appearance in 1910. Maybe investigations like those into the historical-cultural context in which quantum physics developed (Feuer, 1974, and Serwer, 1977) should be extended, with special attention to the whole period leading up to the Great War. Nor is the social history of this time without tremendous upheavals (labour struggles and political protest in Britain, the United States and Continental Europe, intertwined with the various colonial exploits in Africa and so on). To the historian, all these social and cultural events mark an age of dramatic innovation and signal the most acute stage of modernization.

7. Quoted by Prigogine-Stengers (1985), p. 294 from the Einstein-Besso correspondence (1972).

8. Prigogine-Stengers, 1979, Italian trans., 1981, p. 215 (I quote from the Italian edition and translate from this, since it differs in certain respects from the French original and English edition (1985), and at some points represents an integration of these two).

9. It is in connexion with the propagation of signals that a limit is determined to which each observer is subject: the velocity of light in vacuum is the limit-velocity for the propagation of signals of whatever kind; it restricts the region of space that can affect the point at which the observer is positioned.

10. Prigogine-Stengers, 1979. Italian trans., 1981, p. 275; cf. English edition, 1985, pp. 296–7.

11. *Ibid.*, p. 218; cf. English edition, 1985, p. 218 (sic).

12. It is worth recalling the names of the protagonists in this basic change in conception, remembering that they were fully, and combatively, aware of the gnoseo-logical implications of their achievements: see especially Bohr (1958), Pauli (1961), Heisenberg (1959); and see also the correspondence of Pauli with Bohr, Einstein and Heisenberg (1979).

13. On the importance, in the context of contemporary science, of the problem of "reintegrating the observer" in his own description, see Von Foerster (1984). In effect, this "reintegration" runs through all the most significant developments in contemporary physics and biology. Especially as regards quantum mechanics and Heisenberg's uncertainty relations, see D'Espagnat (1971) and (1979); for thermodynamics and the relations between entropy and information, Atlan (1972); for cosmology and the debate about the "anthropic principle" Davies (1982). On the dependence on the observer, which occupies a leading place in certain important notions of contemporary biology (integrone, holone, hierarchic system), see Koestler (1967) and Allen-Starr (1982).

14. Prigogine-Stengers, 1979, Italian trans., 1981, p. 276; cf. English edition, 1985, p. 300.

15. *Ibid.*, 1979, Italian trans., 1981, p. 274.
16. *Ibid.*
17. *Ibid.*, pp. 210 and 276–7.
18. In dealing with the findings of the new biology authors have attempted to systematize their concepts in terms of classical scientific rationality. Out of this, and especially from current theories of evolution, has come a strong reaffirmation of the *historical dimension* (which the interpretation of those findings demands and makes possible). On this debate see Ayala-Dobzhansky (1974), Bendall (1983), and Mayr's recapitulation of evolutionist thought (1982). The new theories emphasize the role played by singular, contingent events that are unrepeatable in evolutionary phenomena. The historical character of evolution and the non-classical nature of its laws have been duly stressed by Dobzhansky (1974), who has also been one of the artificers of the modern synthesis. In effect, if we would understand and penetrate the creative character of evolution in the production of new forms, we must *transcend the classical conceptions of scientific law and causal determination*. This has already been happening in physics for some decades now. The category of necessity is "relativized", *returned to a historical dimension*. The constrictions on evolutionary processes, on their part, take on a historical character, are themselves susceptible of evolution. Emphasizing the historical character of evolution requires that we clearly dissociate ourselves from the concept of prediction and the stranglehold it has: see Stanley (1981). All of which has meant, and continues to mean, that the qualitative change in evolution be brought out; and that this type of change must today be accepted by the scientific community. See Bocchi-Cerruti (1984), pp. 42 *et seq.*
19. Prigogine-Stengers 1979, Italian trans., 1981, p. 282; cf. English edition, 1985, p. 301.
20. In this connexion contemporary epistemological opinion has undergone a radical change. See especially Von Glasersfeld (1979), Watzlawick (1981), Dupuy (1982), Livingstone (1984).
21. Buckle (1869).
22. Hempel (1942).
23. This point of view can be found in Nagel (1952), who recognizes the individual character of the object of the historian's attention. But more attention to the "diversity" of such an object is given in the studies of Gardiner (1952) and Dray (1957). Studies which already take their stand outside the neopositivist perspective.
24. Topolski (1983).
25. These are ancient, by now classic, distinctions: that of Windelband between the nomothetic and idiographic sciences dates back to 1894; that of Rickert between natural science and historical knowledge to 1896–1902; the relation to values in historical knowledge was similarly assumed. Yet these formulations remained influential among historians long beyond the time and the cultural context in which they were advanced.
26. Examples are the works of Croce (1915), (1938).
27. Topolski (1983).
28. There are various examples of this tendency: Rasmussen (1975) proposes a "broader" definition of science, and adds that, in any case, history cannot have recourse to generalizations and laws; it can explain events only by reconstructing them. But Lesnoff (1974), too, thinks that historical explanation must rely on the reconstruction of events. Others, like Gardiner (1952) and Atkinson (1978), have proposed a revision of the nomological deductive model of Hempel and Popper which would make it more suited to historical research; but in their view historiography should be answerable to

laws of a very particular kind. I shall return to these authors for a more detailed analysis of their positions, as well as others.

29. This was the Rickertian "relation to values", clearly distinguished from value judgement; in this context, as in other scientific fields, the latter was to be excluded (Weber, 1904).
30. Compare the remarks of Pietro Rossi (1970), pp. 96. *et seq.*, and those of Kittsteiner (1985).
31. Weber (1903–6) and 1917).
32. Kittsteiner (1985).
33. Prigogine-Stengers 1979, Italian trans., 1981, p. 210. cf. English edition, 1985, pp. 252–3.
34. Engels, 1894, English trans. 1939, New printing, 1966, p. 152.
35. Lombardo Radice (1978), pp. 18 *et seq.*
36. Badaloni (1970), p. 91 (my trans.).
37. Weber (1904) and (1906); compare also Pietro Rossi (1970) pp. 96–98.
38. Hempel (1942).
39. Von Wright (1971).
40. Fain (1970).
41. Atkinson (1978).
42. The first sequence is proposed by Atkinson, the second by Von Wright.
43. Martin (1977).
44. Topolski (1983).
45. The conception of history as the study of single events, understood in the narrowest sense, is emphasized in theorizations like those of Veyne (1971) and Leff (1969). Since the International Congress of Historical Sciences in 1975 we have witnessed a real restoration of traditional procedures; narrative history has come back into its own. Among those taking part in the animated debate on the subject are Thompson (1978), Stone (1979) and Hobsbawm (1980). See also the essays by White (1973) and (1978).
46. In this perspective there may be obvious contradictions. Carr seems to me a case in point. Alongside procedures that are perfectly customary in the narrative reconstruction of historical events and can be found in his research, in his remarks on methodology he wholeheartedly supports causal explanations; but then finds himself compelled to fall back on correctives and adjustments, such as the notion of "hierarchy of cause" or influence of chance (Carr, 1966).
47. Prigogine-Stengers, 1979, Italian trans., 1981, pp. 226–27; cf. English edition, 1985, pp. 223–4.
48. *Ibid.*, p. 232; cf. English edition, 1985, p. 226.
49. *Ibid.*, p. 234; cf. English edition, 1985, p. 228.
50. *Ibid.*, pp. 231–2; cf. English edition,1985, p. 228.
51. *Ibid.*, pp. 233–4; cf. English edition,1985, pp. 228–9.
52. *Ibid.*

CHAPTER 5

Towards a Unified Perspective: Form and Historical Knowledge

In the new perspective the expansion and development of probability theory beyond its terms of application in quantum mechanics marks an important stage. The study of unstable dynamical systems does indeed lead to a genuine association between probability and irreversibility.

Representation of dynamics tends to be substituted by representation of probability which includes the idea of irreversibility. Temporal evolution is no longer determined by the configuration of the type of force—as was the case with the Hamiltonian—but rather by the type of process. But this involves a different mode of determining temporal evolution, without the use of the trajectory, and requires that a "second time", internal to the system, be taken into consideration.

In dynamics time has always been taken as a simple parameter for describing the trajectory. When dealing with unstable systems an operator T can be introduced, giving another meaning to time. On the one hand there is the time associated with the trajectory, which is the time we read on our watch, external time; on the other hand we have internal time, represented by the operator T. Every unstable dynamical system has such a time T referring to the internal state of the system[1].

The internal time may fluctuate and the existence of such fluctuations is represented by new uncertainty relations. "The instability of the motion affords us new conceptual possibilities: the description of evolution in terms of operators acting on the distribution function. . . Entropy can be linked to a new operator that acts on the distribution function. . . There is a close relation between the internal time T and the new operator of entropy M"[2].

In dynamics temporal evolution is described by the Hamiltonian; here a supplementary quantity M enters. This gives rise to new descriptions of the temporal evolution which would be inconceivable in the context of the classical concept.

Now, the expansion of the notion of probability to a point where it is associated with irreversibility, the substitution of representation of dynamics with representation of probability including the idea of irreversibility, the possibilities of distinguishing temporal evolution in new ways—reaching beyond the representation of dynamics that retain the Hamiltonian character of temporal evolution—the introduction of a "second time" internal to the system and able to fluctuate, and, lastly, the admission of a common time direction in a "participatory universe": all these represent new conceptual perspectives.

And all of them combine to determine a new perspective such as invalidates the antagonism between time as pure measure and number (of classical physics) and the time of irreversible phenomena, or rather, time correlated with structure within the chemical and biological processes[3].

By admitting the "time direction" into natural processes, we remove the main impediment to reuniting the natural and human sciences. The persistence of the gulf between them and, indeed, the scissors movement of their progress that created an almost unbridgeable gap between the one and the other type of knowledge, produced a real logical divide between, on the one hand, the concept of possibility in dealing with human actions and, on the other, the determinism of the laws of nature.

But now, as we have seen, the association between probability and irreversibility has at last enabled the burden of theoretical determinism to be shed and replaced with a logic of possibility. "Tomorrow is no longer included in today". The fulfilment of this process will involve admitting into natural and human evolution a sense of history common to both.

There emerges the conception of a "participatory universe", very different from the Newtonian conception of an objectivized and ideally static universe before which man could stand only as an external spectator.

In that conception the notion of time was introduced from outside by man; and, on closer inspection, we find that man's unilateral and quite separate historical dimension was actually justified this ground. This dimension was, of necessity, detached, marginal, intrinsically weak, related to a time which was, in reality, illusory and subjective, a mere convention of measurement.

In the "participatory universe", where man is both actor and spectator and rediscovers a basic continuity in his relationship with nature, his historical dimension coincides with that of nature itself.

This regained common historical dimension of nature's time and man's is reflected in the properties which characterize their conjunction, on the one plane as on the other. We are dealing, in effect, with a number of manifold times, internally correlated with and corresponding to various phenomena. More precisely, the complexity of these times derives from the fact that they can be collocated in a historical dimension (at whatever level of evolution—chemical, biological, social—this type of process unfolds).

The linkage of a historical time cannot but involve reference to the unrepeatability and singularity of the event: it thus enables local, singular, indeed historical relations to be established between the events[4].

But here we reach a new grammar of scientific knowledge, responsive no longer to the language of separation but rather to that of reconnection. This further orientation of the convergences towards a unified perspective opens the way to new relationships between the scientific fields and these, in turn, enrich the latest and most surprising developments in those sciences.

And, indeed, the understanding of living systems requires that we take account of both reversible and irreversible processes. Detailed research into molecules has shown that these systems include both equilibrium processes and far-from-equilibrium ones[5].

The novelty of the significance and the theoretical implications of contemporary biology is most strikingly in evidence in the perspective offered by the physics of irreversible processes. Obviously, the laws describing close-to-equilibrium conditions remain valid; but they do not assist us in analyzing living systems. And, in sum, the entire biosphere, with its living and non-living components, stands in a condition far from equilibrium[6].

Indeed, the importance of establishing relationships between physics, chemistry and biology—as these relationships became possible at certain points in the development of these sciences, and in the perspective of their relinkage—is better understood from a study of phenomena far from equilibrium.

Molecular biology affords the best example. This discipline has shown how non-linear reactions (relatively rare but by no means absent from the inorganic world) are almost the rule in living systems. Indeed, the transmission and utilization of genetic information is governed by a non-linear mechanism (a feedback loop).

Thus we can get a clearer glimpse of the prospect of a logic of evolution characterized by the existence—in far-from-equilibrium states—of self-organizing processes of various types, processes of which life itself represents the fullest expression.

But the self-organization of matter involves, in varying degree, a *cognitive property* (either more simply implicit and closed, or more pronounced and open, according to the degree and level of evolution at which it stands).

When by shifting the system away from equilibrium we reach a threshold of instability, we arrive at a "bifurcation point". If we consider possible further bifurcations, we see how the system can lead to various possible results, which will be more and more numerous: by moving from stable to unstable behaviors, each time we reach a region of instability, the system "chooses" its possible future. If we bear in mind that the state reached at each bifurcation depends on the previous *history* of the system, we shall get some idea of the actual historical path it travels[7]

Obviously the fluctuations around a bifurcation are wider than usual. The system then begins to "choose" between various possibilities (which at that point are highly indeterminate).

Close to equilibrium a structure can be altered only when its environment changes; far from equilibrium the fluctuations enable the differences present in the environment to be used in order to produce a different structure: it is as though in such conditions matter were able to perceive its environment and distinguish small differences which would have no importance at equilibrium. The behavior of matter in such conditions is characterized by the ability to perceive and communicate[8]. It is as though, on moving away from equilibrium, matter showed intelligence.

These new conceptual views, arising out of the most recent developments in the natural sciences, mark a further step in the reverse trend of thought with respect to the relations between the sciences imposed by the Newtonian construction: a trend which has assumed increasing importance since the first break with the old conspectus that was signalled by relativity theory and quantum physics.

With regard to the problem of the scientific status of history and the other sciences of society, we have seen what implications may arise as we pass from a deterministic model to a probabilistic one. We have seen how, with the association between probability and irreversibility, the gap between traditionally separate scientific typologies can be further closed.

Comparison of methodologies thus made possible with respect to the development of the sciences and the new relations between them can now hardly be avoided, when we confront the latest results in the study of far-from-equilibrium states.

This study has been the object of a highly interdisciplinary research such as to range beyond the traditional boundaries that demarcate the various departments of scientific investigation. Many of these studies have dealt with the diversity of levels of evolution and have addressed themselves to comparing and relating the phenomena referrable to these different levels. Just as frequent is the reference to the feature of self-organization assumed by evolutionary change in conditions far from equilibrium.

Even a summary review of the studies that have contributed to outline the new perspectives of unification would not be easy[9].

A first and essential reference must be made to the researches into the thermodynamics of non-equilibrium and to their implications for, and comparative extensions into, biophysics, on the more general aspects of the phenomena of self-organization, leading ultimately to consideration of biological and social phenomena: Katchalsky-Curran (1965), Nicolis-Prigogine (1977), Prigogine-Stengers (1979). Secondly we must refer to the dynamical theory of systems in mathematics with the studies of Thom (1972), Zeeman (1977) and, more recently, Abraham-Shaw (1984). It is well known that the influence of these studies has ranged far beyond theoretical mathematics and made itself felt in the scientific debate over the last ten years. Specific studies, with considerable interdisciplinary implications, have also contributed to the intersection between problems in physics, chemistry and biology, such as those of Morowitz (1968), Eigen-Schuster (1979), Epstein-Kustin-De Kepper-Orban (1983). To these may be added others on the dynamics of systems and processes of self-organization, such as Haken (1978), (1980) and Jantsch (1975), (1980). Another important interdisciplinary contribution has come from the cosmologist Chaisson (1981).

Then again, in the new perspective of relinkage a central role has been played by studies of evolutionary biology, beginning with the now classic Jantsch-Waddington (1976), Stanley (1975), Eldredge-Gould (1972), (1977), Gould (1977a), (1977b) and (1980), and including the special or more recent contributions from Ager (1973), Csányi (1982) and Denton (1985).

Elsewhere in this book I have emphasized the importance of the new theories of evolution beyond the more narrowly biological context, as well as the problems concerning the socio-cultural evolution of man. I have also quoted liberally from the contributions coming from the other side: the biology of knowledge, neurophysiology, the theories of cognition, psychology.

Yet we cannot take account of the new perspective of relinkage without emphasizing, once again, the essential role played in this connexion by "transversal" disciplines: cybernetics, information theory and, above all, general systems theory. With reference to the last-named the reader is reminded of the work of von Bertalanffy (1968), Weiss (1971), Laszlo (1973) and Pattee (1973).

The theoretical implications of the study of non-equilibrium systems demand a rigorous rethinking of the relationships between the sciences, in a perspective from which we cannot exclude the human sciences—even if we wished to do so. For human beings are themselves far-from-equilibrium systems, in as much as they are living beings and organizers of societies; and then, as we have seen, these systems—whether they have to do with natural or social evolution—can be collocated in a historical dimension of time, a real time which is no longer a simple parameter of movement but a measure of internal evolution for a non-equilibrium world.

It was thus a premiss of this development that the new concepts, introduced into the study of far-from-equilibrium systems, should not only allow but almost necessitate a relationship between the trends of evolutionary processes and those of socio-historical processes. One might even speak of analogy, were it not preferable and much more meaningful to concentrate on the theoretical problems that emerge from the confrontation, and the gains in methodology that accrue thereby.

By confronting the results of the studies of systems far from equilibrium in the area of the natural sciences with results obtained in the area of the socio-historical sciences, we can pinpoint properties and characteristics whose value assumes a more general significance, far beyond the specific field. It will thus be useful to develop this kind of confrontation, at least as regards certain of the most significant points.

In effect, the behaviors of far-from-equilibrium systems show an evolutionary pattern characterized by significant properties. The process passes through successive choices between the possibilities available at the points of bifurcation. At this juncture, the further the system (pre-biotic, biological or social) lies from equilibrium, the more able it appears to perceive and communicate—in other words, to know and to interact with the environment. Which is to say, it evolves by producing new structures, through a process of self-organization showing a property of cognition.

We have noted how, each time the system "chooses", it does so on the basis of its previous history; so that through a succession of such choices—passing from stable to unstable regions each time a bifurcation is reached—a genuine historical path is traced.

The historical character of the evolutionary process thus sketched is confirmed and further evidentiated by the fact that, when the "choice" is made, the laws we can use to describe the behavior of the system can no longer be of the universal kind; they must become of a specific type, or, more precisely, they depend on "the type of non-linearity involved when we cross the boundary between equilibrium and nonequilibrium"[10].

This is a clear manifestation of the connexion between the character of specificity and singularity, proper to a historical trend such as that connoting a similar process of evolution, through bifurcations, and the matrix itself of that specificity and singularity, inherent in the logic of self-organization.

We have seen how a system can modify itself, both as a result of events taking place within it, and by influences from outside, when innovations occur following uncontrollable events.

Under those conditions the relations established in the process of change in the system have pronounced historical character. Thus, in the case of modifications produced by internal events, it is important to note how—in the presence of abnormally wide fluctuations—relations may be set up between events that were formerly different from, and independent of, one another and at a certain point become interconnected. In other cases there may be local modifications, occurring in a particular region but able at a certain point to propagate and affect the entire system. Again, when innovations (which may be mutations or technological modifications) are introduced, the new constituents, taking part in the processes of the system, may give rise to transformations that compete with the previous modes of functioning of the system: if these transformations prevail, the system will function according to the new patterns, thereby acquiring a new morphology.

Actually we are dealing here with mechanisms (or strategies) of change whose properties extend beyond the environment or type of phenomena in which they can from time to time be detected. It is not simply a question of such mechanisms' being detectable in certain phenomena at very different levels from one another (from glycolysis to mucilaginous moulds, from the anthill to the industrial revolution), however strikingly important this may be.

Rather, such mechanisms show a more general paradigmaticity of significance and describe paths of evolution that may be found on various planes. In effect, they trace a general evolutionary outline whose historical character comprises a multiplicity of phenomena and evolutionary levels to which they can be ascribed—from the most elementary through the most complex biological processes to the socio-cultural processes of humanity.

By shifting its attention from the equilibrium of forces to the irreversibility of processes, the new physics has enabled other sciences of irreversible processes, social as well as natural, to be set alongside one another in paradigmatic terms.

This has led to a unification of perspective in the configuration of a process of evolution that can be studied in an interdisciplinary way. The new perspective is reflected in the thermodynamics of non-equilibrium and in dynamical systems theory, together with physical cosmology, evolutionary chemistry, macro-evolutionary biology and the sciences of complexity; it has led to a completely new conception of evolution. ". . . evolution takes place step by step and from one level to another, alternating stages of determination and indetermination, unfolding processes of self-conservation through change and processes of reorganization. . . . "[11].

If referred to the socio-historical plane, the study of this evolutionary dynamics enables us to detect the mechanisms of change and innovation which characterize processes that we may define, in historiographic terms, as those of "crisis and transformation". In historical research such processes assume the greatest importance, and particularly when we analyze the passage from previous structures to new ones.

More generally speaking, historical evolution itself can be represented in terms of the process of crisis and transformation. Historically, considering the long term, we can remark that this process of crisis and transformation is itself, too, characterized by a series of times or regions of stability and instability: the new structure affirms itself prior to moving to a new instability which marks the threshold of a new possible future.

The socio-historical events that take place at this threshold are analogous to events in the world of nature occurring at far-from-equilibrium conditions—in the second case these may be metabolic reactions, items of animal behavior or whatever.

Thus in the case of a change as in that of a technological innovation, the fluctuations may be especially wide, such that the possibilities, which include the "choice" that is finally made, will be particularly various and undetermined. So that the *morphological* novelty introduced by the "choice" with respect to its initial possibilities will be evident: the choice represents an *acquisition*.

Therefore we cannot take a reductionist view of this step either as necessary or as mere chance. To put it another way, the acquisition resulting from the "choice" represents not the pure and simple predominance of the most suitable morphological connotation—on the basis of necessity or chance or both together—but rather the working out of this connotation.

This is obviously very important if we consider the mechanisms of change with reference to evolution, both natural and socio-historical.

If we now approach these problems with an eye to the evolutionary process, we must analyze what the mechanisms of adaptation are in a system far from equilibrium, like those which may occur at any level in the process.

In a far-from-equilibrium system we can detect a sort of "interface" of the relationship between interior and exterior; or to put it another way, a dual sensitivity towards the fluctuations produced by the internal events and towards the fluctuations arriving from the external environment, which is itself in a state of flux. In these conditions a system would appear to be capable of "translating" the fluxes by which it is fed; this very ability to "translate" enables it to "organize adaptation"[12].

Thus it is not merely subject to adaptation, but actually performs an active function of adaptation, i.e. even with respect to the mechanisms of adaptation, it behaves according to a logic of self-organization. And this, as we have seen, implies a perceptive-communicative ability, i.e. a cognitive property. Any possibility of interpreting or reinterpreting these mechanisms in a deterministic way is ruled out by the fact that this perceptive-communicative ability is also reflected in the mechanisms of adaptation: we cannot reduce the adaptation to mechanisms of selection understood deterministically. Selection resumes its original meaning of choice—choice meaning acquisition and innovatory elaboration, the more innovatory the further it stands from equilibrium.

In this aspect, too, of the evolutionary process there emerges—consistent with all the other properties which describe its progress in far-from-equilibrium states—the characteristic of self-organization of matter at all levels; and this self-organization implies the possession of a cognitive property.

Thus the basic character of the nature-knowledge nexus at each degree and level of the evolutionary process, natural and socio-historical, is confirmed.

The intelligence of matter ceases to be a mere postulate or intuition, a somewhat metaphorical expression to describe the results achieved in certain sectors in physics, chemistry and biology[13]. This intelligence may be inscribed and described in an evolutionary process, in a *history* of evolution, unfolding on the basis of the nature-knowledge nexus. And it is only to an evolutionary process thus conceived that we can belong, unless we insist on claiming that we are alien to it—an absurd contradiction.

In order to site the nature-knowledge nexus in theory, as a basic constituent of the evolutionary process, a new scientific paradigm must be adopted: that of *form*, representing the active synthesis in which that nexus is expressed.

The properties characterizing the mechanisms of change in a far-from-equilibrium system can be summed in the synthesis of the form.

The theoretical model of this synthesis can thus be presented as a comprehensive model, i.e. one able to explicate the various possible trends in the course of a change, by whichever mechanism this happens among those that we have seen as capable of stimulating the transition to a new structure.

Actually, the synthesis of the form does not simply represent a result, a mere datum as the outcome of a process. The form is the form itself of that process: i.e. it represents the property by virtue of which that process can occur. We are dealing, moreover, with an extremely flexible model (as we shall understand better later on), since it represents changes of various entity and extent: that is to say, it represents them at each degree and level of the evolutionary process in which they occur.

Note, moreover, that this is not a theoretical model superimposed on the results of research: the form interprets the results and re-orients them, transposing itself within the research and becoming a constitutive moment. The synthesis of the form can thus be verified at every moment, on the basis of the afore-mentioned results, and directly answers to the arrangement they require.

We should therefore ask ourselves, and verify, whether and how the form is able to represent and assume in itself the properties and mechanisms of change that we have already seen to characterize an evolutionary process by successive bifurcations—this each time the threshold of such a change is reached, in far-from-equilibrium conditions.

The changes occurring in such conditions have three basic characteristics.

Firstly, we have seen how, with respect to the especially indeterminate possibilities which appear at the critical threshold of change, this change effectively represents an *acquisition* and involves a *morphological* novelty. The form is, indeed, the form of this acquisition: not form *qua* result, not a mere datum, but rather a "forming" form governing the change—an internal "ratio" of a structure in evolution.

Secondly, in the most complete way, the form accounts for another fundamental characteristic of the change; this can be verified in any system that moves from a previous structure to a new one, in far-from-equilibrium conditions. In such conditions, and at any level of the evolutionary process, the system behaves as a whole: i.e. it shows a consistency such that, whatever kind of relation is established within it or in the interaction with what is external to it and that initiates the mechanism of change, this will affect the mode of functioning of the whole system—or rather, its organization.

At molecular level, for example, the system assumes a structure as if each molecule had "informed itself" as to the overall state of the system[14]. The same problem arises at the level of the cell: how does differentiation come about among the cells that go to make up various tissues and organs? Even when we adduce the useful enough notion of morphogenetic field, how does a cell manage to "recognize" its position? And the examples could be multiplied, moving to other levels. We could enquire what logic governs the apparently, but only apparently, casual movements of termites in the construction of an anthill[15]. On the one hand, a deterministic explanation which attempts to isolate a causal factor will certainly be reductive. On the other hand, the temptation to fall back on a teleological explanation sidesteps the problem and also represents an arbitrary simplification. What no reductive explanation can account for is the shift from one order to another: for instance, from molecular activity to the supramolecular order of cells; which also implies a different dimensionality. This kind of difficulty has been frequently and clearly pointed out[16].

Elsewhere in this book we have already seen how problems of this type create a need for theory, for a theory of organization; and how, in the new conceptual framework of the study of far-from-equilibrium systems, this must be interpreted in terms of self-organization. We have seen, moreover, how the process of self-organization implies a cognitive property which can indeed account for that ability to perceive and communicate that characterizes the behavior of matter in far-from-equilibrium states.

Then, and only then, can we deal with what would seem to be the most important problem: the ability to "recognize", which appears to be, and indeed is, intimately linked with biological activity, whether molecular, cellular or at some other level. Such an ability can be well represented in the synthesis of form, by virtue of which it is at one and the same time a *performing* and an *understanding*. And this seems to me the intimate reason behind that overall consistency which appears in the behavior of the system as a whole.

And here, once more, we become aware of the need for an unambiguously *anti-reductionist* explanatory model. But this is not a generic need—i.e. along the lines of the classical anti-reductionist formulation, tending simply to establish a rather different hierarchy between the whole and the parts. The model must, rather, correspond to a more "comprehensive" conception—one capable of giving specific account of the role played by the parts in relation to the behavior of the system as a whole.

The ability of the form to resume, also in this important perspective, the properties of the mechanism that define the system as a whole, indicates the modern anti-reductionist character of such a model: it indicates, i.e. how the form constitutes a model entirely corresponding to the anti-reductionist requirement arising from the study of far-from-equilibrium systems.

Thirdly, in the most consistent and flexible way, the form shows itself able to represent the features of historicity in change in a process of evolution by successive bifurcations.

In this connexion it must be emphasized how each of the mechanisms of change referred to above leads to changes characterized by *singularity*. Singularity that cannot but be defined historically. In fact it should be noted that the outcome of the said change, or better, the structure which announces itself in a system that has reached the point of bifurcation, will depend on the previous activity of the system, on its history. Moreover, the singularity itself of the new structure represents the uniqueness of the actual outcome, as against other possibilities encountered at the threshold of change. This outcome, in turn, will constitute a historical precedent—that is to say, a determinate and irreversible trait in the subsequent course of the structure.

But this singularity belongs to the evolutionary process, is written into it. Thus it cannot be detached or isolated as an event considered in and for itself alone, whose uniqueness is conceived idiographically—as in the case in the more traditional conceptions of history[17]. Rather, it must be linked with a process of self-organization which, in far-from-equilibrium conditions, unfolds along a genuine historical path.

The form will thus be the form of that singularity, but it will be so inasmuch as it pertains to an evolutionary process proceeding in terms of self-organization. It will be the form of that process: not the simple morphology of a result, but the property by which that result occurs. It therefore represents, at one and the same time, the singularity of the new structure and the process leading up to it. So that we can say that the form does not represent the outcome of a change, the datum, but rather it is the form of that change.

Let us now consider all three features of change that occur in far-from-equilibrium conditions that have been dealt with above, viz: the fact that change involves an *acquisition* expressed in the form itself of the change; the fact that in the course of the change the system behaves as a consistent whole; the fact that the change takes on the character of historical singularity that cannot be isolated from the process that underlies it. All three of these features effectively describe a mutation proceeding in terms of self-organization.

The form is the active synthesis representing this process of self-organization. Moreover, the form resumes in itself the particular properties and features of the change occurring in the process. And, as we have seen, these are historical features describing the evolutionary change in historical terms—whether that change takes place in biological or in socio-cultural processes.

It is the historical character itself of the process, detectable at two different levels, the natural and the social, that is the most important and significant datum. The paradigm of form must be verified in correspondence with that chracter.

Form appears as a consistent and flexible model, able to represent the singularity of the change and, at the same time, the coherence of the totality in which it occurs, the incorporation of the change in a general process of evolution, and the new element, the specificity that it involves each time.

The change can also vary in size and extent, both at natural and social level: sometimes it will mark a decisive stage, of particular significance in the historical path traced; at other times it will constitute a more partial or more limited acquisition. Thus the form can also account for the intermediate steps in the process, as well as for the more striking, more critical moments. But in the one instance as in the other, the change will in any case be significant, it will mark some step in an overall process.

Form thus enables us to describe the evolutionary process as a self-organization—of the living being and the social being; this self-organization proceeds by continual approximations, through acquisitions which may be only partial or relative, seeking and *discovering,* from time to time, its paths among the various possibilities.

So the evolutionary process appears as a long, non-linear historical-explanatory journey—as befits a nature that knows itself through evolution and describes a historical path through that evolution.

If in this way form shows itself able to interpret and represent the data and the paths of the evolutionary process—as they can each be reconstructed and verified—we shall be able to consider form as a new scientific paradigm for the study of evolution, both natural and socio-social, in a new perspective of linkage between the two areas of science.

But, as we have seen, this relinkage involves a redefinition of the very bases of the socio-historical sciences.

To relate historical knowledge with the process of general evolution, and to recognise a fundamental historical-exploratory character in the changes that characterize this process, is possible only if we conceive the process as one of self-organization—like that which unfolds in far-from-equilibrium states at the prebiotic, biological or socio-cultural level.

Thus the urgent question emerges: how should we interpret scientific results in relation to this essential point?

It is well known that in the opinion of some authorities biological organization can be explained as pure and simple selection and accumulation of the various favourable mutations: the organization of man then become a highly unlikely result; and this path may lead to the conclusion that the very existence of man is a result so improbable as to be equivalent, in practice, to mere chance.

This kind of conclusion will of necessity be reached if one tries—as Monod did—to arrange the results of contemporary biology in a classical conceptual context[18].

Referring instead to a quite different context—that afforded by the physics of irreversibility and by the new convergences with chemistry, biology and human sciences which can be postulated on this basis—we can tackle the problems of organization that arise in the study of far-from-equilibrium systems, in all their complexity and having regard to their interdisciplinary relationships. We shall thus be in a position to understand how, in such conditions, even tiny deviations and individual behaviors can be subject to amplifications, assuming roles in provoking change in the system such that its functioning can only be configured in terms of organization. It is to that organization that the consistency of behavior of the system as a whole and of the mechanisms of its transformation must be referred.

If, therefore, we wish to give an adequate interpretation and a closer analysis of the results of contemporary biology, and, more generally, of the problems posed by the study of far-from-equilibrium systems, we cannot do without a theory of organization.

In the new conceptual context the interpretation of those results and problems leads us to conceive of evolution as a process of self-organization. The validity of this can be verified only if it brings a gain in research methodology.

As regards the problem posed in this connexion, i.e. that of a re-linkage between the socio-historical sciences and the natural sciences, which enables us to bridge the logical and methodological divide imposed by the Galilean paradigm—there can be no doubt that, with respect to what we have dealt with hitherto, the verification mentioned above will be entirely positive.

And not only as regards the problem of methodology, highly important as that is, but even more with respect to the consequences implied by the change in perspective in terms of the very conception of history.

On the other hand, it is clear that in a quite different, indeed opposite, perspective, like that indicated by Monod, historical knowledge would end by finding itself in contrast, and in a marginal state, analogous to the position it occupied in the traditional system of scientific relationships.

Indeed, in reaffirming a classical conceptual framework, we would only reinvest history with the old antinomies—above all, with that of chance versus necessity. Historians have had long experience of the weight of this burden.

In connexion with history, to state that a certain event is accidental obviously means that no explanation of it can be given. The fact that this event is interwoven with others, to which some degree of cause is instead attributed, means that only at this second level—that of causal concatenations—can we find elements that will prove the explanation. The historian may also have recourse to ingenious combinations in order to locate the chance element in a chain of necessities[19]. But in the end, however he may have interwoven the twin terms of chance and necessity, in explaining the facts he has chosen as significant he must come to terms with the causal chain. Admitting chance into the process will not absolve him from the duty of giving a casual explanation[20].

If we return to a concept belonging to classical physics where the peculiar features of the initial conditions are contrasted with the deterministic universality of the laws of evolution, we shall only reconsign history to the choice between the labor of Sisyphus of climbing the slope of that determinism and the resignation to stay at the bottom.

But the question is not limited to logic and methodology: an entire conception of man and of history is repostulated, and thus the "humanistic" prejudice is readvanced.

In Monod man is the result of chance and necessity, won from chance to the realm of the immutable laws that govern nature, so that he returns to that external position, abstract and solipsist, where classical science has placed him. And now, as then, the conception of his exceptional, if fortuitous, particularity sets him apart from nature, in a condition that is marginal even with respect to the very sciences that concern him.

While the "humanistic" prejudice prevails, man can even rejoice in this diversity-estrangement, though it actually diminishes him. The condition of marginality to which he will return must also apply to his entire history.

Quite otherwise is the position of the historian in a really new conceptual framework that enables him to deal with processes in far-from-equilibrium systems—which are, by definition, the ones he studies. He can incorporate such systems in an evolutionary process of a historical-exploratory kind which unfolds by self-organization.

The historian can approach a theoretical problem of this kind through the paradigm of form, in order to interpret the results of research and re-orient them in the historiographic perspective; as in the relinkage between this and other sectors of science concerned with far-from-equilibrium states, he must refer to a quite different concept of order: order through fluctuations. Thus the historian will be enabled to distinguish between the states of a system in which individual initiative, single events and particular situations have small significance or no influence at all, and other states in which facts and situations of the same entity may, instead, play a decisive role in provoking change. In this second case the historian will be able to understand the really innovative character assumed by actions and ideas in particular circumstances: which occur when the actions and ideas can avail themselves of the same type of non-linearity as characterized the system in its previous states, in order to provoke changes[21].

Essentially, then, for historical phenomena as for natural phenomena, it happens that, in the presence of new conditions, an identical type of non-linearity that characterizes a particular structure can profoundly alter that structure, giving rise to a new one. In such conditions the traditional dichotomy between chance and necessity fails.

Hence, especially with regard to historical research, it must be clear how much is gained from the fact that in this connexion, as at other levels of evolution, we can no longer view an organization or a working regimen as merely necessary; nor does the fact that they are characterized by fluctuations mean that they are merely arbitrary[22].

The theoretical implications of a way of proceeding like this are considerable: the "choice" that appears at the point of bifurcation may include an element that we should traditionally define as chance but which, on closer examination, shows no arbitrary character.

The "choice" thus assumes a character that cannot be properly defined in terms of traditional logic. It is as though the result of a choice continued to retain its initial premisses even while representing something new—an acquisition, indeed, with respect to those premisses.

The new element brought by the choice, with respect to its own possibilities, *includes the sense of those manifold possibilities.*

But it is the synthesis itself of the form that represents a "choice" of this kind, in the most consistent way: the form actually includes what it tends towards; and since what it tends towards is already included in the form of that tendency (or better, the form already contains what it tends towards), what occurs, or can occur, does so according to that very form-tendency.

The form, therefore, *qua* active synthesis of a historical-evolutionary process unfolding through self-organization, will describe the change that characterizes the said process in terms that reach beyond chance and necessity. Thus the form constitutes a new theoretical model for historical knowledge—both in this specific field and at other levels of evolution.

This would seem to offer a considerable gain in perspective, in logic as well as in methodology; as an explicative model the form shows itself to be extremely flexible and comprehensive in the analysis of mechanisms of change, and one very much capable of verification.

NOTES TO CHAPTER 5

1. Prigogine-Stengers, 1979, Italian trans. 1981, pp. 250–53. (I quote from the Italian edition and translate from this, since it differs in certain respects from the French original and the English edition, and at some points represents an integration of these two).
2. *Ibid.*, p. 252 (my translation).
3. On the problems concerning the conception of time in classical physics and time as a measure of evolutions within a world of non-equilibrium, see: Grünbaum (1964), Schrödinger (1956), Cohen (1962), Toraldo di Francia (1976), Romano (1981).
4. On the ancestry of this problem in the history of scientific thought, see the remarks of Serres (1977) p. 136 et seq. For a review of the notion of time in a specifically historical context, among other Pomian (1977).
5. The examples that may be adduced in this connexion come from the researches of Goldbeter-Nicolis (1976), Goldbeter-Caplan (1976), Hess-Goldbeter-Lefever (1978), Goldbeter-Segel (1977). Other equally significant studies have been performed by Blumenthal-Changeux-Lefever (1970), Lefever-Deneubourg (1975). On the latest results and indications of research in this direction see Cramer-Freist (1978).
6. On the importance of referring to these conditions in the analysis of a living system and on the various levels of such analysis—microscopic and macroscopic—see Cramer (1979) and (1984).
7. Prigogine-Stengers, 1979, Italian ed. 1981, pp. 160 and 164.

8. "Matter at equilibrium is dull. The further one goes away from equilibrium the more intelligent matter becomes": it is surely significant that a statement like this is repeated and fully dealt with by a biochemist like Cramer (1986). In effect he explicitly advances a new concept of matter.

9. I can, however, avail myself of Ervin Laszlo's systematic reconnaissance, carried out from a point of view I fully share: Laszlo (1986).

10. Prigogine-Stengers, 1979, English edition, 1985, p. 181.

11. Laszlo (1986), p. 22 (my translation).

12. Prigogine-Stengers, 1979, Italian ed. 1981, p. 176. These behaviors have been observed in nature by Arnold-Horsthemke-Lefever (1978), Horsthemke (1980) and others. For more general remarks on the notion of adaptation see Lewontin (1977).

13. I refer, in particular to studies such as those of Prigogine-Lefever (1975), Richter (1979), Cramer-Freist (1987), Schuster (1984), Hess-Markus (1984).

14. Prigogine-Stengers, 1979, Italian ed. 1981, p. 165.

15. Cf. Deneubourg (1977) and Bruinsma (1977).

16. Weiss (1969), Waddington (1975) and Jacob (1983).

17. The idiographic nature of historical study, in so far as it is centered round the description of individual facts, begins with the classical formulations of Windelband back in 1894 and arrives at the most recent restatement by Veyne (1971) of an analogous mode of understanding the singularity of the historical event. For a more general treatment of this problem see Topolski (1983).

18. Monod (1970).

19. One example of this is Trotsky's reference to the phenomenon of refraction in his autobiography: "The entire historical process is a refraction of the historical law through the accidental. In the language of biology, one might say that the historical law is realized through the natural selection of accidents" (Trotsky, 1930, pp. 421–22).

20. Carr (1961), Chap. IV.

21. Prigogine-Stengers, 1979, Italian ed. 1981, p. 187.

22. *Ibid.*, p. 188.

PART TWO

Morphological analysis applied to historical phenomena

CHAPTER 6

The Dynamics Of Change
In Historical Process

AN EXAMPLE OF COMPARATIVE ANALYSIS: SOCIAL
CONFLICT IN A TIME OF "MODERNIZATION".

We have considered, from a theoretical point of view, the properties that
inhere in the mechanisms of change and transformation in far-
from-equilibrium systems. We have seen how these mechanisms—in
conditions of dynamic instability—lead to morphological change in a
system: the previous working regime is replaced by a new one and this
leads to the formation of a new structure. We have shown the properties
that enable us to analyse these mechanisms in the dynamics of complex
systems both in the area of nature and in that of society. The present
chapter has something of the character of an appendix. In it I propose to
give an example of this kind of analysis by referring to a particular process of crisis and socio-historical transformation. The fact that the mechanisms of change and transformation—in conditions of dynamical
instability—have been studied almost exclusively with reference to natural phenomena does not mean that our investigation will consist of a
simple transposition, by analogy, of this sort of analysis from one context to another.

I believe that the theoretical basis that has emerged from the foregoing
chapters provides the terms necessary to perform the analysis at the different levels of reality, without attenuating or losing sight of the specificity of those levels and thus of their autonomy that justifies the
particular area of investigation.

The example I shall develop will, rather, provide an opportunity to
show how and to what extent this kind of analysis, even in our particular
sector, may offer useful contributions for both the specific and the general aspects of the elaboration and the working-out of the theory.

Lastly, my study will aim to verify what advantages, what gains for knowledge may accrue from the formulation I propose, as compared to the more usual models of explanation in history. To this end, the process of crisis and transformation that I shall examine is one of the most important in modern times. It has to do with the birth of a social conflict at mass level. This conflict soon became systematic and irresolvable. I refer to the social unrest that accompanied the beginnings of industrialization and a more marked development towards capitalism in the rural areas.

The phenomenon will be dealt with by morphological analysis of social change. This will enable us to examine how it contributed, in combination with other factors, to the passage from one structure of social organisation to another, new one. For it seems to me that over and above the specific aims they were "subjectively" pursuing, over and above the realization of those aims (often only partial, sometimes non-existent), the movements I describe did combine to set in motion a new social dynamic—one that was to lead to nothing less than the birth of the modern structure of society.

And this shift of perspective may lead to an abrupt transcendence of the more traditional formulations of the history of the working class movement.

As is well known, these formulations formerly cramped themselves by fixing too narrowly on the goals and strategies that the workers' movements envisaged at trade union and political level. So much of the history of labor movements has largely focused on the decisions and tendencies of the leaders, for the strategic options they made; almost as though the fate of the movements depended solely on how right they were and how effective. In practice, this meant seeing things from the same point of view as those protagonists of history, rubber-stamping their pre-eminently political-subjective terms of evaluation. This lack of detachment can be seen in certain features of a strain of "militant" history-writing which is still too much concerned with justifying or criticising ideological and political decision in order to establish its own ascendancy. The result is a flattening of the history of workers' movements: this in favor of a predominantly political history, indeed, in the even narrower sense of tendencies and policies of leaders. And this has led to a neglect of the more truly social dynamics that ultimately involve the "grass roots" of the movement. Contemporary historiography has critically surveyed this type of formulation and its adverse judgement may by now be taken as a *fait accompli*[1].

There is, however, one aspect which I wish specially to highlight. To re-establish, in its totality, the distinction between the political-subjective aims that the various organized movements proposed from time to time, and once again to measure, as we can and must, the actual effect of their actions and achievements, gives scope for broader and deeper analysis of the function "objectively" performed by them as components of a new social dynamics.

The analysis that would appear feasible in this area concerns, in effect, not only results and achievements to be evaluated in terms of relations of force (that underwent more or less lasting changes); it also concerns, and far more nearly, the new forms that came into being through the demands advanced in the conflicts and the way those conflicts were organized. The new way of looking at social relationships and how they were expressed (even if only partially), the new associations— economic, trades union, political: these things made up a novel morphology of social organization, which at more general level contributed to delineate not unimportant aspects of modern social formation.

The problem I propose to deal with is absolutely relevant here: the identification of a new area of analysis. And the necessity and utility of a reformulation of this kind can easily be evidentiated when one considers the notable problems still arising from this subject.

In the specific case the problems do not differ from those of a more general kind that I have already dealt with in the theoretical part of this work. Here, too, we see the unsatisfactory nature of explanatory models that, basically, belong to the causal type (however far this typology is corrected, attenuated or qualified).

In the case of the social conflict at the beginning of industrialization and the fuller establishment of agraria capitalism, the problem appears in the following terms. On the one side, we have historians who continue to emphasize the transformations in technique and production, considering them as determining factors in explaining the birth of workers' and peasants' resistance movements and the forms they assumed: in other words, primary elements in the process of formation of those classes and their trades union and political consciousness. Here we are obviously not dealing with the old economistic formulations, nor with narrowly deterministic ones. Nevertheless, however hard one may seek to treat as central the reference to the "class", to its direct experience of the contradictions in the production process, the analysis continues to revolve round consideration of the "objective" factors connected with the amount and degree of transformations in the process of production[2].

On the other side stand those historians who are critical of this formulation and propose to stand it on its head. Their attention goes to the "subjective" factors in the formation of the working class, taking quite different elements into consideration: elements of culture, mentality, so cial behaviour. But this read may lead to a reduction, in the opposite direction to that of the first school, to a similar banalization of the problem of social conflict. The risk, which has even been admitted, is that the treatment may be watered down into social history in "kind"[3].

In my opinion the dynamics of change should be dealt with directly and in specific terms. In the context of social transformation like in any other, it should be possible to examine in what conditions and in what ways a complex dynamical system may undergo a change in its structure, passing from one working regime to another.

The casual type of model of explanation will thus be replaced by the reference to the mechanisms of change and transformation that we have observed to characterize the dynamics of complex systems. For the purposes of our analysis we can usefully refer to two such mechanisms:

a) the establishment of relations between events that were formerly independent of one another;

b) the change in working regime of a system due to the introduction of an innovation (technological or otherwise);

b2) in this case although the innovation competes with the system's previous modes of operation, it may "combine" with these in the new relations of transformation that set off the change.

(This latter trend, which can be seen as a variant or a specific trend of mechanism (b), is recurrent and can be clearly observed in the socio-historical context).

In historical analysis it may come about that mechanisms (a) and (b) intervene in the same critical process of transformation, and in certain cases they may actually combine. It would be interesting to discover if and how this occurs in other areas of scientific investigation.

My hypothesis, in fine, is that reference to these mechanisms of change and transformation in the analysis of the historical process I propose to study here brings with it the following advantages:

1) it enables a particularly effective analysis of how and why a new type of social conflict developed—in the presence of industrialization and a fuller establishment of agrarian capitalism;

2) it enables us to go beyond all causal formulations (however attenuated or qualified by conditions) to explain the relation between the two orders of phenomena—economic and political, while not losing sight of the correlations between them, and indeed identifying them with greater precision;

3) it is capable of explaining the composite and contradictory characters, or even the elusive ones, that those movements show, as compared with the schemes of interpretation that seek to ascribe them

narrowly to the degree of economic transformation, or to purely political desires and strategies;

4) it enables comparison between the beginnings of these conflicts in
contexts that differ greatly one from another and exist in different
periods;

5) it gives us the elements necessary to explain outcomes other than the
conflict in a dynamic of change unfolding in other conditions of dynamic instability; and in this case, too, it offers useful instruments
for comparative analysis.

The Italian case

The outbreak of mass social conflict, that was soon to become implacable and capable of assuming new forms of organization, began in the
country areas of the lower Po Valley in the 1880s.

The struggle was quick to consolidate and spread. It involved other
land workers from categories and areas different from those who were in
at the start. Soon enough it also involved quite broad strata of non-
agricultural labourers: these were above all textile workers and people
employed in other sectors not yet completely alienated from their peasant
background. And within a decade the movement spread to embrace several strata of the urban proletariat, in large and medium-sized towns,
mostly in the Centre-North of Italy.

The largest section of the urban proletariat and the job categories were
the first in the struggle and the first to organize themselves; they belonged rather to the traditional and mixed city "artisans and working
people" than to the new factory proletariat. And there were further
strikes in the 1890s among a numerous sector of workers in the building
and allied trades, these, too, belonging to a traditional area and labor
typology that was not agricultural.

The social composition and the developments of this new workers'
movement in Italy thus exhibited a highly elusive character; indeed, in
many respects a decidedly contradictory one, when we set it in any
scheme of interpretation that seeks to base the new social conflict on the
consistency and the more or less advanced degree of capitalist transformation that was under way.

According to this scheme, the proletariat of the new factory system
should have been the first to contest the state of things and provide leadership for the movement. Other labor sectors could follow them, depending on the degree of capitalist transformation in the area to which they
belonged.

As for land workers, their role was to follow faithfully in the wake of the industrial workers, not to lead them; and they could follow only in so far as they were conscious of the process of capitalist development that was in progress.

The contradiction has by no means been neglected by Italian historians. Solutions have been sought in a supposed "political precocity" of the Italian workers' and peasants' movement. This precocity has been seen as having something of the character of a surrogate with respect to the partialness, the inconsistencies and the contradictions of the capitalist development in Italy[4].

For some time this sort of interpretation seemed to credit the case of Italy with being special, even perhaps anomalous; but in the long run it has turned out to be unsatisfactory. In my view, if the Italian workers' and peasants' movement at the turn of the nineteenth century showed considerable vigor at social level and experienced considerable political radicalization, it is these very elements that demand explanation; otherwise, one risks falling into a real tautology. In any case, even if we would re-direct the search for determining factors from the economic-objective plane to the political-subjective plane, we cannot forget that the schemes that prevailed in the Second International, and in the Italian socialist movement, would have involved strategic options and alliances that would have resulted in an inverse trend and a logic of development quite different from those that actually obtained. In fine, I would argue that, in an analysis that sticks closer to the facts, one must give up any sort of "apriorism" in the search for "determining factors": that is to say, one must abandon schemes that are in any way causal.

Shifting the analysis to another plane—in the terms we have outlined above—we can see how, in the conditions of dynamic instability that characterised part of Italian society, the transcendence of the critical limit of that instability was due to a specific mechanism of change, capable of operating in conditions like this. Actually this rupture was brought about by the establishment of relations between events that were formerly independent of each other and occurred in different parts of the world.

What events? Firstly, the growth of vast new cereal-producing areas, above all in the Americas. Side by side with these, other types of agricultural production took on increasing importance, among them rice and textile crops, typical of southern and eastern Asia. Then there was the large-scale establishment of steam navigation with a resulting intensification of ocean transport and a sharp fall in the cost of ship hire, which also became convenient for carrying larger quantities of less precious goods. Then again, the Suez Canal was cut, so shortening and considerably facilitating commercial communication with the countries of South East Asia. Lastly we should add the high degree of commercialization and the increased scale of exchange of agricultural produce, especially cereals, in the traditional European production areas, together with the rising demand for raw materials for the textile industry. Nor should we neglect the great expansion of railway systems that also fostered the exchange and transport of products all over the continent of Europe.

The relations established between these various events combined to create a new regime of competition and a sharp fall in the prices of the main agricultural products, and this led to a vast crisis in European agriculture.

However, this crisis set off very different processes in the various agricultural contexts and even within single countries (as occurred in Italian agriculture). Nor is it a question here of how far the economic-quantitative element in the crisis affected matters, but rather of the specific conditions in which that combination of events operated as a mechanism of change and led to the transcendance of the critical threshold of social instability and the start of a new dynamic.

In the contest of the rural areas of the lower Po Valley, where the new social conflict had its beginnings, the process of crisis and transformation involved a rupture in complex economic and social equilibria that had already reached limit values of saturation and stability.

Throughout the first half of the nineteenth century those areas had witnessed an increase in the groups of migrant workers who moved between the areas of wet crops (rice paddy fields) and dry crops (wheat-maize and wheat-flax rotation). For it was the complementary nature of the various crops that enabled workers to find jobs at different times of the year, according to the season of the year and the particular type of crop. By travelling wherever they could pick up employment, they managed to accumulate the minimum number of working days that would enable them to subsist[5].

In that particular conjuncture, the landowners were concerned to change their crops in order to defend themselves against the fall in prices of certain crops, and this led to a drastic reduction of rice paddy fields. In many cases these were replaced by rice fields that could be turned over to grass leys. And the traditional wheat-maize or wheat-flax rotation was cut down in favor of fields providing animal fodder. In various

areas, especially the low-lying ones, the wet crops were wholly neglected; and when it became convenient to drain them, quite new crops
were introduced, such as sugar beet. In every one of these cases the aim
and the result was to make a drastic cut in the labor force required. The
net outcome of which was that the amount of annual working days reck-
oned per unit of productive area plummeted[6].

But the most important, perhaps most traumatic effect was the complete disintegration of that complementarity between the two crop culti-
vations—so long established and guaranteeing at least a margin of
subsistence to an already fluctuating, unstable population of agricultural
labourers. In addition to that, the instability extended to share-croppers
and smallholders, who were evicted. The fluctuations of the labour market and the margins of subsistence had already reached the limits of instability; under extreme pressures from the relations that led to the
agrarian crisis and to a certain type of response to that crisis, it was not
long before they passed the critical threshold.

The urgent and continual need for jobs immediately produced conflict.
It was, of course, contradictory to the logic and the aims of the partial
transformation of production, which itself had been carried out in response to crisis and involved a further contraction in the work force employed. At the same time these transformations also provided the motive
for resistance or rather for a search and demand for other job opportunities in the reclaimed lands or those under new crops. The resistance on
the part of casual and other farm laborers had a defensive character: its
essence was none other than a search and demand for jobs. But this demand became incompatible with the new conditions. The first ''movements'' consisted of groups of laborers who went into the fields in
spontaneous processions, or wandered from one field to another
in search of work. This merely involved more groups than before making
demonstrations of their demands which had become more insistent and
less susceptible of being satisfied than in the previous situation. Before
it became intentional, the character of conflict, of rupture with what
had gone before, of these processions, assemblies, movements, demonstrations, demands and so on, derived from the same logic of dynamic
change; but it was only a matter of time before that logic was overturned, in the nature of things.

In those conditions resistance could only repeat itself systematically,
season after season, occasion after occasion; it could only become more
insistent and widespread. And in this way it took on the value of conflict, more and more pronounced and evident, becoming and even more
organized and systematic ''movement''. That this could happen was

thanks to the fact that the new social dynamic took off on the basis of the same type of non-linearity that had characterized the previous working regime of the system.

The first associations of laborers were resistance organizations. The real content of their objectives was the demand for work and/or regulation and remuneration of work that, in the new circumstances, should guarantee them a subsistence.

The element of conflict and the irrevocable character taken on by the movement, the element of novelty and alternativeness of the new union organizations, and their break with previous social and political relationships—these quickly came to constitute elements of a new dynamic, irreversible and destined to establish itself and modify the previous working regime of the system.

Having passed the critical value, the threshold of instability, the extension of the fluctuations required a new organisation and a new logic that pervaded the system in its totality. And this is most important if we would explain the principal features of the subsequent developments of the resistance movement, developments of far greater entity than their beginning might have led one to expect.

The first upheavals in the farm laborers' struggle were felt in 1882 and '83 in the provinces of Mantua and Cremona, with some stirrings south of the Po, in the provinces of Piacenza, Parma and Bologna.

But the "break" came in 1884–1885, with the movement known as "la boje" (dialect for "it's boiling"): this highly significant figure of speech was used by labourers to refer to the phenomenon of "convection"[7] that took place in their protest movement. Which rapidly spread: it began in the harvest season of 1884 among labourers in the Polesine, who demanded more work and higher rates, and swiftly involved labor in the province of Ferrara on the opposite side of the Po. Then in the year following it spread to the provinces of Mantua and Cremona and also further north to those of Milan, including the region of Monza, to Como and Lecco. South of the Po the movement found support in the provinces of Parma, Reggio and again Ferrara[8].

In this opening stage the main nucleus was represented by casual laborers, but other land workers were also involved. And over the following years unrest among the textile workers, alongside the struggle of agrarian labor, was to bring to the fore once again certain areas in those parts of northern Italy we have just mentioned.

Soon enough a further characteristic of this type of conflict became clear: this was its tendency to move in waves that affected now one area, now another. But, simultaneously, certain epicentres appeared, where the protest was more persistent and more systematic, returning in more or less each season.

In the second half of the 1880s strikes by laborers became more and more frequent in the flat lands of Bologna and Ferrara. From these we can see how the epicentres of the struggle were located precisely in those areas having much sharper features of that non-linearity that we mentioned above[9].

Those same years witnessed a more limited unrest among other workers in other areas: for example, the tenant farmers in north Milan province and the Como region; but these were also of significance as they involved areas where agriculture and textile manufacture (especially silk)[10] were strongly interdependent.

In the first half of the 1890s the protest movement of the land workers spread and strenghened its organization. From the epicentres of the Bologna and Ferrara plains it extended into the province of Ravenna, on one side, and the plains of western Emilia, on the other. On the opposite bank of the Po the struggle was resumed in Cremona, in systematic fashion[11].

Above all, these years saw the movement grow in political and union organization. The *leghe di resistenza* (of laborers in the Bologna fenland as of tenant farmers in Cremona province) represented new structures. In the epicentres of the Emilian plain various associations of land workers and some cooperatives had already been formed in the previous years. In the early eighteen-nineties the phenomenon became more widespread, took on sharper outlines and became more closely wedded to the political activities of the socialists[12].

In 1897–98 massive strikes broke out in the Emilian plain and again in Mantua province. In close connexion with these an intensive and more widespread political-trades union agitation was undertaken. So that, following the suppressions of '98 and '99, and in spite of them, the year 1901 saw the founding in Bologna of the *Federazione nazionale dei lavoratori della terra* (National Land Workers' Federation): this was the largest Italian trades union on national scale, with 152,022 members, of which 68,586 in Emilia Romagna alone[13].

In that same year several strikes occurred throughout Emilia Romagna, spreading across the Po into Cremona and Mantua provinces; here two thirds of the farm labor force were involved in the *leghe di resistenza*[14].

By now the interweaving of resistance movement and union organization constituted a new reality, running like so many threads through that social fabric.

And in similar circumstances other areas and other land workers were affected, if only in partial and limited ways, by this process of crisis and transformation that had firstly and most directly involved the casual labourers of the lower Po Valley. Nor was this all. By the same mechanism of change and the same relations set up between events as described above, workers in the textile trades, traditionally linked with agriculture (flax, cotton, silk), were also affected.

Leaving aside the case of flax, which belonged in the picture of complementarity of crops analyzed above (since it, too, was mostly grown in the lower Po Valley), another type of complementarity must be described: this is represented by the widespread persistence, in many areas, of outwork or cottage industry in spinning and weaving. This activity (in other contexts, too) existed side by side with agricultural labor and complemented it. Even where real manufacturing work was involved, this could not entirely be detached from the framework of complementarity with agricultural labor (we find examples of worker-peasants or a primitive division of labor within the family or rural community); here, too, we find complementarities combining to guarantee the bases for subsistence. In addition, the sharp drop in agricultural prices, including flax, cotton and silk[15], set off a crisis in the spinning and weaving sector that ultimately assumed a trend similar to the one I have described in agriculture production. Here, too, there were partial conversions and restructurings in production, but the modes and systems of employment of labor, and its exploitation, far from undergoing any essential change remained the same as before, wherever possible.

For textiles the repercussions against the effects of instability and the onset of a new social dynamic were, partly, of the same kind as those already experienced by the agricultural workers, and partly more specific, as we shall see later on.

In effect the relations established between events previously independent of one another, as we saw at the start of the process, affected the entire working regime based on complementarity between different sectors and production typologies. The social area involved was tending to expand and the repercussions were destined to make themselves felt far beyond their starting-point.

It was the complementary activities that guaranteed subsistence to the land workers in the Po Valley and others who were in similar circumstances; and this subsistence was now threatened from both the quantitative and qualitative points of view. The question was not simply one of the amount of working days per year or the wages that could be earned

by a single worker, or his family, by employment in the various agricultural and non-agricultural jobs.

The division of labor, the distinction of roles and tasks in the family
structure, as in the tissue of relationships in the local community, were
based on the ways in which these activities could be reconciled and the
customary nature of their connexion; so that equilibria had been
achieved, even if only minimal ones, within the systems of life.

To be sure, this was no "harmonious" world, with workers at peace
with their circumstances. I have already stressed that a limit-condition
had been reached, characterized by an endemic precariousness and a tension already pressing towards the critical threshold of social instability.
For these very reasons the crossing of that threshold had repercussions
that at a certain point could no longer be controlled.

All of which explains how the social breakdown and the onset of conflict of a new kind came to occur first of all in the special context of the
Po Valley, and subsequently spread rapidly beyond its initial epicentres
to the adjacent areas, until it involved those sectors and typologies of
industrial labor nearest agricultural production.

However, these repercussions and developments in a broader social
framework remained congruent and can easily be referred back to the
same mechanism of change and transformation that had set in motion the
new dynamic. So that it can clearly be seen how, in these circumstances,
not only the reasons for the conflict, but also its very forms and organization resembled one another; which favored the consolidation and expansion of the struggle.

By and large, we have here a dynamic of social change that was taking on the connotations of a new morphology, connotations even more
widespread and clearly detectable. It is also important to emphasize that
the new dynamic—set in motion once the critical threshold of instability
had been passed—had a very precise relation to the type of non-linearity
that characterized the previous set-up.

Thus it was the possibility and the capacity of the socialists to relate
their own action to the type of non-linearity that characterized the social
conditions in which they intervened that enabled the socialists to play an
effective role in the organization of the movement at the beginning. And
if the times, modes and efficacy of their political initiative were not
identical between one context and another, this was due, or mainly so, to
the same set of reasons.

In this way, in the lower Po Valley, the Socialist groups, which had
initially enjoyed very little influence, were able to link themselves to the
movement, interpreting its needs and developing along with it. These

were fairly heterogeneous groups as regards social and political-cultural formation. They centered round prestigious leaders and local newspapers, or were organized in branches of the recently consituted socialist party. Hitherto their work had been mainly propaganda and had rarely involved any broader links with the population. It was characterized by strictly political relationships at national or international level rather than by an extensive and persistent presence among the working classes themselves.

The new development they experienced could only come about following the onset of a mass conflict in the rural areas, whose needs they understood and could interpret for the very reasons we have seen above. And at that point their political-organizational activity became efficient and incisive beyond all comparison with their previous record. In other words, we find a real connexion between political action and aims among those socialist groups and the mass movement that had come into being in the country regions.

Hitherto our analysis has aimed to explain the character of the movements and the motives that inspired them. We must now see how the social conflict developed through its successive phases, and how its further expansion embraced other strata of the working classes and involved different social realities. These latter differ, to varying extents, from those we have dealt with up to now and thus require a differentiated analysis. And, in effect, the trend of the process of crisis and transformation in the industrial sector involves reference to quite another mechanism of change.

Generally, speaking, as in the Italian case, this mechanism has to do with the introduction of a new production technique: this enters into the processes of the previous working regime of the system and, by combining with them, may change them.

But a mechanism of this sort cannot be understood in a reductive, deterministic way. And that we have here something quite different from a deterministic relation between the introduction of new technology and the change that occurs in the system, is shown by the fact that the innovation may, or may not, alter the equilibria in such a way as to modify the system's working regime.

In any case, in historical analysis one can scarcely ascribe the quality of cause to an innovation in production technique, taken in and for itself. Firstly, although the innovation is available or already applied in other contexts, in many cases it may not be adopted at all (even where its introduction would be economically viable). Secondly, a different, even if less apt, solution may be adopted. And even if the innovation is introduced, it will not necessarily make itself felt. The already existing set-up of the system may be of a structural stability such as to push it into the background[16]. So that one needs to examine in what circumstances it will work so as to bring about a change in working regime of the system. In the case of Italy we must bear in mind a basic feature, due to the fact that industry too, as a whole, remained for a long time on the brink of large scale capitalist transformation: this transformation was a more or less partial one and the interests and possibilities in this connexion differed considerably from sector to sector.

As we know, the formation of a real industrial base in Italy came quite late—between the last years of the XIX century and the early years of the XX. In international terms this corresponded to the second stage in the process of industrialization. Which meant that a very high degree of technical and productive concentration was required; together with a new type of relation with finance capital and economic policy[17].

As against that, the difference between technical and production standards and market relations that resulted from the world-wide-process of industrialization made it hard for Italian industry as a whole to enter the market and survive competiton. And as a result, Italian industrialists, like the agricultural producers before them, sought and obtained a strongly protectionist policy from the government[18].

By no great paradox, the result was to buoy up the strongest groups and the most advanced sectors having greater concentration; whereas the more traditional sectors and technical-productive typologies were very badly hit by the crisis.

This second category of industries was by far the most numerous and the one most vulnerable to the crisis. Here innovations were introduced in order to restructure, but only in so far as restructuring was possible and could be used to combat the crisis.

In effect this restructuring was limited in two senses: on the one hand it involved drastic reduction in the manpower originally employed in the particular sector or type production.

In social terms this resulted in increased unemployment and the replacement of skilled labor with cheaper manpower; the consequences were thus analogous to those that ensued in the transformations in agriculture and textiles.

On the other hand, technical innovation was superimposed on the old modes of employment and exploitation of the work force, all the more so when these old methods were typical of tradition-bound industries; so that the new tended to combine with the old, giving rise to intolerable as well as contradictory situations.

The most frequent case, for example, was where the worker continued to be paid on piecework, while his labor now depended on a new relationship with machines and a new organization of the production cycle. Thus he was deprived of any possibility of controlling his own work, yet was paid or penalized as though fully responsible for it. His working situation became absolutely contradictory, with a grossly unfair calculation of his wages, and it was thus that his situation became quite intolerable.

And note that we have here something more than a mere economic datum: as I mentioned above in connexion with the break in the economic-social tissue of a certain kind of agriculture, the contradiction also affected the socio-cultural side.

In the case of the industrial workers the culture of their calling was enfeebled, emptied of its content, together with all that that culture had meant in terms of social identity. The upheavals in the labor market had direct repercussions on the structure of the family and the modes of life. The widening fluctuations, the passing of the critical thresold of instability, affected everything that had been consistent in the previous structure. And here, too, the worker was compelled to defend himself, to struggle, to resist.

The mechanism of change was different, and the workers' protests assumed different forms and specific contents. These had to do with the quantitative and qualitative terms of his labour, i.e. with wages and norms (pay rates, working hours, factory regulations).

But even in the case of the industrial workers, the protest was initially a defensive one, yet wholly in antithesis to the logic and interests of restructuring: a protest that was also destined to assume the trend of an irreconcilable opposition, strongly organized and systematically conducted.

So that the new social dynamic expressed in this conflict also appeared in direct relation to the type of non-linearity featured in the previous composition of the system.

In the Italian context this dynamic, affecting the struggles of labor, opened up another front in the social conflict in addition to that of the land and textile workers. Clearly, however, this mechanism of change also managed to superimpose itself on the other one that we described above (a mechanism due to relations between different events) as the dynamic of conflict began to take effect. This certainly occurred in the textile sector which suffered in two respects: from the crisis in the regime of complementarity between agricultural and non-agricultural production, and from the logic of innovations and partial restructuring going on in industry.

It therefore comes as no surprise to find that the textile workers alone accounted for 30% of those on strike in industry throughout the period when industrial strikes mushroomed, between 1892–1900[19]. And the dual aspect of the crisis undergone by the textile workers was well reflected in the timetables, and locations of their protests and the reasons that motivated them.

The first signs of conflict in this sector can be detected at the close of the 1880's. The outwork weavers in Como province, the silk weavers there and in the districts round Lecco and Varese began to agitate, mostly against the reduction in work offered, the dismissals and the lower piece-rates and wages, all of which the entrepreneurs used as the most immediate means of reacting to the crisis[20].

But it was from 1893 onwards that the conflict in the textile sector took on more consistency and more precise features. In June of that year agitation began in the yarn and cotton factories in the province of Bergamo and rapidly spread to other textile areas, especially silk, involving the provinces of Cremona, Brescia and again Como[21]. The demand for fairer wages and the resistance to further exploitation are witness to how the owners were beginning to act arbitrarily and ruthlessly in order to meet the crisis.

It was soon evident what was going on: those typical partial restructurings were being made in the working cycle so as to cut down the labour force and dequalify and more fully exploit the manpower retained. A very precise logic was at work: partial innovations were introduced in technique and organization, but keeping as far as possible to the old criteria in the regulation of working relations (that basically followed a pattern of piecework). The new did not replace the old, rather it was combined with it, thus considerably worsening the situation of the workers and giving rise to blatant contradictions. This can easily be seen in the struggles of the textile workers in 1896 and the years following. Innovations in production technique and organization were translated into cuts in pay rates and/or increases in penalties and wage stoppages, for faulty work or methods of working that were no longer under the control of the worker. Who thus found himself having to pay doubly for his increased exploitation.

This was the reason why the Cremona yarn spinners began a long agitation through August and September 1896. Similar reasons underlay the strikes in the cotton and wool factories in '96 and '97. Especially harsh was the struggle of the weavers at Biella in 1901[22].

More generally, the worker's loss of control over his own work—for which he was nonetheless still responsible according to the old piecework system (which was disguised in various ways)—and the resulting decisions by the owner as to the rates and methods of payment: these were the motives behind various recurring strikes in textiles in the second half of the '90s and the opening years of the XX century[23].

As against that, we should not forget that, alongside the "restructurings", other owners reacted more simply and drastically to the crisis, with mass dismissals and heavy wage cuts. The latter were the object of an attempt at resistance by the cotton workers in Turin and Bergamo provinces and the area of Monza, with strikes between 1895–98, and by the woollen workers in the Val di Mosso in '96–'97[24].

But the crunch was also felt by other workers in cottage industry, operating in different sectors that were nevertheless still bound up with agricultural production. Their protests were very determined and involved a very large number of strikers (as was the case, for example, of the female straw-plaiters); and these things combined to make a considerable contribution to the overall pattern of "industrial" strikes in the 1890s[25].

Closer analysis shows that this pattern has more elements in common with the struggles in the primary sector than might be abstractly supposed. There were certain points of contact between these different productive sectors and typologies: even the conflict in the industrial sectors furthest detached from agriculture showed an important element of protest against unemployment and the worsening of the workers' situation of instability.

A characteristic of this kind can be seen in the protests of workers in the building and allied trades: they were the second largest group of strikers (17%) in industry in this period[26].

But here, too, we must stress the existing relationship between the crisis of unemployment that affected this sector between the end of the early 'nineties and the crisis that hit the agricultural laborers. Their ranks had actually produced a large portion of the work force employed in house-building and the building of infrastructures in large and medium-sized towns, mostly in central-north Italy during the period of expansion that preceded ours. Internal movement between these two areas of the labor market was then considered customary. Now unemployment in agriculture was added to the recession in building.

There was large-scale and violent agitation by masons, builder's labourers and allied workers in 1891, in Milan, Rome and other towns, and again in 1897 spreading over a large part of urban Italy, above all in the central-north region[27]. These protests against unemployment and unstable working conditions in various respects recalled the typologies of the protests of the land workers (tendency to go in waves, systematic refusal to accept settlements offered and rapid extension of the strike front).

These points of contact and convergence, together with the spread of trade union and political organization (not lacking) from the protests of one sector to those of another, fostered the coming together of associations of workers and associations of land laborers or tenant farmers in a horizontal type of organization that operated in both urban and rural areas. This was the *camera del lavoro*.

The *camere del lavoro* were originally supported by local administrations and had the function of mediating in the labor market. But thanks to the common content of struggle against unemployment and instability of working conditions the function substantially altered, and they too became organizations for trade union action.

For a variety of reasons the protest by the industrial workers initially wore a defensive character (as well as an implacable one), like the earlier protest by the casual laborers.

In combination with the elements of contiguity established between the movements among the casual laborers, textile workers and workers in other sectors, this helped the *leghe di resistenza* to imitate and propagate the kind of organization originally obtaining among the casual laborers.

With the passing of time and the persistence of protests in the various sectors, as well as the continual expansion of the protest front, there was a trend towards convergence at union and political level. But times, modes, relationships could not be taken for granted. Differences remained and were not easy to settle. Although to unify objectives were frequently vitiated by the underlying motives for divergence and opposition between the various movements.

The general picture of the workers' and peasants' movements in Italy, despite the ever more forceful leadership of the Socialist Party, was distinguished by the variety of its composition, by its different origins, by the non-linearities of its specific dynamics, even in the year following our period. For this reason historians have always found it very difficult to interpret; and, as compared to the customary schemes of interpretation, it has appeared highly contradictory and elusive.

But if we focus our analysis on the irreducible specificity of the social dynamic, we can understand the "real movement", and thus once more examine the economic and political trends without reduction or flattening out in any way.

The analysis here proposed, even though conducted purely by way of example, enables us to focus on the different dynamics of social instability that affected the Italian context, and on their relations to events that occurred outside this context although exerting an influence on it. The dynamic instabilities led to differing outcomes, once the critical threshold was passed. These make up the overall picture (non-linear) of social change in Italy produced by the process of crisis and transformation that we have examined.

The morphologies of change can easily be identified and compared with those occurring in other contexts. In each case we can establish the limits and content of the outcome, and we can study how they eventually converged to produce a further evolution.

And with all the limits we have mentioned, the results of the protest in the rural areas and industry of Italy at the turn of the XIX century were to prove lasting. They gave rise to a trade union and political movement that developed up till the First World War and, immediately afterwards underwent its greatest expansion in the struggles of the "biennio rosso", before its defeat at the hands of fascism.

Initially, both the agricultural workers' and industrial workers' movements had a defensive, sectional character. Soon enough, for the reasons we have seen, they assumed a systematic and implacable trend. They immediately exhibited the character—not episodic but permanent—of the new protest and social dynamics they set in motion. Nor is this something to be taken for granted as due to a simple deterministic sequence.

It is, therefore, important to enquire: what was the dynamic that enabled the movement to take hold so rapidly and so immediately consolidate? The phonomenon is one of crosscatalysis, similar to the one essential to the mode of evolving of far-from-equilibrium systems, as can be observed in nature. We have here, in fact, a catalytic reaction in which the different products or groups of products operate as catalysts towards each other for their synthesis[28].

As we saw, the resistance on the part of both agricultural laborers and industrial workers was defensive but had a content of irreconcilability; was, indeed, in antithesis to the modes and logic of agricultural transformation or industrial restructuring. So that it assumed the trend of a struggle systematically repropounded and in the same terms: the workers would not desist, the owners and industrialists would not accept the claims. The content of resistance that characterized the struggle and was also reflected in its form of organization (i.e. its product) acted as catalyst for the resistance and organization of the other side (and vice versa). Associations of land laborers and industrial workers were more and more systematically opposed by organizations of landowners and industrialists who sometimes took the same name of "leagues" and responded to the same need to organize resistance, in their turn.

A trend of this kind can be entirely assimilated to a cross catalysis. By this process the logic and aims that determined the actions of both sides became more and more precise. Not only did they consolidate and take shape in permanent organizations, representing new and irreversible facts reflecting the new morphology, but they were also translated into genuine social and political strategies. We have seen how these strategies corresponded to the terms of a new social dynamic.

Thus the entire previous working regime of the system was contested; the various kinds of paternalism of welfare bodies and political bosses, the ways and motives of mediation and coexistence, consolidated in the previous social set-up, and all the organizations and insitutions that represented them—these all broke down or reoriented themselves. The political struggle itself shifted away from the more traditional antagonisms and old divisions no longer held. In the national political framework the mode of representation and mediation between constituted interests was altered, at least in the formal sense.

These were, of course, partial conquests, contradictory results, that marked the onset of a process of crisis and transformation that was still far from over. Several other factors were to make themselves felt, new mechanisms of change were to spring into action. Above all, the war and the post-war crisis were to extend the fluctuations. But even when we look at the historical developments following our period and the climactic stages of the process, the analysis here proposed has an interesting feature; it enables us to identify the initial, and irreversible, elements in the formation, the first stage of a new type of social complexity.

Over and above the specific objectives of the organized workers' and peasants' movement, and the long term aims, which provided the Socialist Party with its *raison d'être,* the protests and their early successes marked the beginning of a new social morphology.

Let us first consider the development of the trade unions. This happened horizontally, with the affiliation of the *leghe di resistenza* of the various categories and sectors of workers to the *camere del lavoro.* But, parallel with this, the *leghe* also converged in vertical organizations, such as the craft federations that represented the workers in a particular sector also at a national scale. Secondly, we must look at the new fabric of society that was being woven together, especially in certain areas, by means of the close relations between trade union organizations, cooperatives, socialist and democratic political clubs, "red" municipal governments. Lastly, the more strictly political aspects of the experience and activity of the organized parts of the working class; party militancy, election constests, new relationships and criteria of political representation and delegation. These were the first elements of an organization, or self-organization, at the lower end of society; some of them autonomous or alternative elements, but all combining to form modern mass society.

However embryonic those realities, however obstructed their developments in the future, however contradictory and contradicted the achievements they sought and attained, for as much as that "possible future" desired by the workers diverged considerably from the future that emerged in the course of the subsequent bifurcations—nonetheless, the process cannot be considered as in vain, historically speaking. In the sequence of bifurcations that followed, it contributed to outline an irreversible perspective.

Viewed and analyzed in these terms, a process of crisis and transformation enables us to give a historical valuation of more partial and more limited changes, as also of the relation to a dynamic, vaster, more complex, unpredictable, that affected those changes and to which they themselves contributed.

By morphological analysis of change—following the methodological rules I have sought to exemplify—one can identify the forms of that change, at the various stages and passages of its self-construction. These should not be understood as phases or results of change—which would make for a forced interpretation or arbitrary simplification. Rather, they are the forms that shape the change itself. The analysis must be of a dynamic in action, i.e. the specific trends of an ongoing process. An analysis of the kind done here does not take us simply to the conclusions of the process, with a final balance sheet drawn up in narrow and often misleading terms; almost as if—to continue with our example—an entire cycle of social conflict could be judged merely on the strength of the final outcome.

The kind of analysis of which I have given an example enables us to "comprehend" the various paths traced as part of a more general process of social transformation, and the meaning of a process of this sort does not lie "outside" but, rather, inheres in the self-constructing of that process.

The English case

At this point I must underline another, no less important aspect. An analysis of the social dynamic of the type performed in the foregoing section provides useful elements for a comparative analysis, since it is able to show the morphologies of the change in question.

The effects of mechanisms of change similar to those in the Italian case can be found in previous stages of capitalist transformation and industrialization in other national contexts. Let us take the case of England at the close of the XVIII century and the start of the XIX, i.e. at the dawn of the process of industrialization considered as a whole.

Now, according to the classical schemes of interpretation, a comparative analysis of the motives and modes of social protest in a "latecomer", like Italy, and a country like England, where industrialization and the establishment of capitalism in the rural areas arrived a century earlier, is not viable. Or, even if we limit ourselves to partial comparisons, the case of England is viewed as a term of reference, in some sense exemplary, with respect to which the events in countries where capitalism developed later would seem, for that very reason, to be characterized by contradictions, inconsistencies, anomalies.

In practice, the first of the two terms is invested with the value of a model of capitalist development. This, not only because England was the first country where that development took place on a large scale and anticipating others but also because—rightly or wrongly—a special consistency has been ascribed to capitalist transformation in England. This consistency has been referred both to the terms of economic development, and to the framework of social relations. Historians have thought to find an exemplary organicity of relationship between primary and secondary sectors in the establishment of the capitalist mode of production. As regards the work force, too, the one that was liberated from agriculture is supposed to have been largely absorbed by the new industry. Lastly, the English bourgeoisie has been invested with a specially consistent role at economic-social level and in the history of English politics.

The English road to capitalist transformation has thus become the "classic" one and has assumed the status of a model. It would take too long here to examine how and why this sort of picture has been built up, so that the transformation of English society in this period is made, as it were, to pay the price for its primogeniture.

Suffice it to show that all this has led to two gross oversimplifications. On the one hand, the history of England in this period has, in some sense, been forced into corresponding to the model that so many analyses have sought to evidentiate. And this has led historians to neglect several aspects that contradict or elude the schematic picture that has been built up.

On the other hand, this type of formulation has ended by giving prime importance to the more strictly economic factors of the transformation:

so that, compared with other components of the process, these factors have assumed the value of determinants.

Ultimately this dual simplification has strongly influenced the analysis of the social conflict, with the result that the latter has been all too neatly explained as the mere outcome of the development of capitalism, and closely depending on it.

As against that, an analysis bent on studying the dynamics of social change, their specific mechanisms and morphological importance, enables us to formulate the comparison in much more fruitful terms.

The case of England can therefore be seen as the first onset, partial, contradictory, slow—as much so as in any other contexts—of a more general process of crisis and transformation affecting the transition from the previous form of society to the modern one. In this process the stablishment of capitalistic relationships of production obviously represents the salient trait, but it cannot be viewed as the totality of the process, as totally determining its outcome; it cannot, to any extent, be assumed deterministically. It is incorrect and misleading to suppose—in this context, or, indeed to a larger degree here than in others—a deterministic relationship between the transformations in technique and production in English agriculture and industry, on the one hand, and the social protest, on the other, at the turn of the XVIII century.

The times, the modes, the protagonists of the social protest in England of that period are as little linear as they are highly complex and elude the classical schemes of interpretation, the direct, causal relationships that historians have sought to establish, for example, between the spread of power looms and the decay in working conditions of the weavers, in cottage industry and traditional manufacture, or to take another example—between the enclosures and the protest movements on the part of agricultural laborers: these may appear logical and congruent in the abstract, but they are much less so when we come to analyze what actually happened. We have only to examine the problem of timetable to find ourselves at once in the thick of inconsistencies and difficulties of interpretation.

Starting with the problem of periodization will show us, indeed, how hard it is to get "order", to interpret the phases and trends of the social protest in England between the end of the XVIII and the start of the XIX century, if we rely on deductive schemes. Not only will those schemes reveal their insufficiency, but we shall see how many problems still remain unsolved.

And how could it be otherwise? In the country that first witnessed industrialization and the fullest establishment of capitalism in the rural areas, that process, far from being especially consistent and complete, was instead slow, tortuous, obstructed. Although the agents of transformation were powerful, the process met with no little resistance. Even if we focus on the aspect of most interest to the present study, we see how the working classes were able to resist, able to show a capacity to respond, not only on the economic level, but also on the socio-cultural one. Their reaction was a strong one and had the means and the time to deploy itself, interacting in no merely passive fashion with the content and the logic of the transformation.

The political arena, above all in the first stage of the protest, saw a substantial defeat of the workers' and labourers' resistance, but was itself affected by the backlash in ways that would emerge more clearly later on.

Superimposing highly schematic categories of interpretation on the trends that actually occurred has led to an abstract classification or a mannerist representation of the stages and convulsions of the social protest.

Cases in point are not only Luddism, but also the "Captain Swing" riots and several other "episodes" of the social conflict. Studies of these phenomena have often missed the connexions between them, neglecting sufficiently to analyse their relationships with the more general aspects of the social dynamic then at work. The acute stages or significant emergencies of a more general process of crisis and transformation have been reduced to rebellions, "Jacqueries", machine-breaking "riots"; other episodes have been declassed to the level of corporative movements, merely in defence of ancient statutes and working regulations.

In the new reformulation I propose here, however, the hypothesis of interpretation and periodization aims to identify the beginning of social protest on a mass basis; the social protest is especially significant as it marks the onset of a new social dynamic, in the struggles that included the Luddite movement, but it cannot be reduced to a mere movement or at least not in the ways in which Luddism is commonly understood. In effect, that cycle of protest extends beyond the more narrow arc of Luddism, with elements of continuity that can be detected in the subsequent development of the process.

In certain respects our periodization coincides with that adopted by Thompson, who sees the years 1811–13 as a watershed, from which a new conflict arose that was to last till the regulating of the working day in 1847[29]. My interpretation of that turning-point, however, differs from Thompson's. The question is not merely one of distinguishing between protests that appealed to the past and those capable of looking towards the future. According to my reformulation, that turning-point can be understood only in terms of the onset of a new social dynamic.

In this perspective we can examine and interpret more precisely the motives behind the outbreak of a new type of social protest in early nineteenth century England; we can single out the specific trends and, at the same time, identify the elements most useful for comparing this process of crisis and transformation with other, similar ones. From this point of view, the methodology I propose will enable some highly significant findings.

For example, we may note the strategic importance, in the case of England, as of Italy, of the equilibria of complementarity, not only economic but social and cultural, that characterized local communities. Among other things, these equilibria depended on the particular contiguity of agriculture and production of non-agricultural goods, above all textiles, in cottage industry or the traditional manufactures. And as in the Italian case so in England, when these equilibria were upset, the critical threshold of instability was crossed.

Now, if the historical conditions of rural society, or cottage industry and traditional manufacturing were so different in the contexts, it is interesting to note how the mechanisms of change that set off a new social protest were, nevertheless, of the same kind. In the English case, as we have already seen in the Italian one, the first mechanism of change to be remarked is that arising out of the establishment of relations between different events.

First of all, there were the Napoleonic Wars, that had already wrought considerable changes in the economic situation and in society; to which changes must be added the naval blockade of the Continent. On the other side, there was the reaction of the United States to the "Orders in Council" of the British Government, regarding neutral countries. This involved complete suspension of trade between the United States and Britain. And the effect was especially serious on the import-export cycle of cotton (American raw material against British manufactured goods); the cycle was already essential to the process of industrialization in England and was to continue to be so. Add to this, moreover, a series of very poor harvests from 1809 to 1812, with a rise in corn prices of 150–180%; this coming in a domestic market for agricultural produce that had already been subject to extensive changes from the traditional pattern of trading. And we must further add a change in consumption that had been in process since the late eighteenth century and particularly affected textile goods.

The combination of these various factors determined an extension of the fluctuations that was to lead to the passing of the critical threshold. Indeed, a new and irremediable blow was dealt at that regime of complementarity whose importance for the effects of the equilibria—not merely economic, but also social—we have already stressed, for the situation in England as for the Italian situation in the previous section. For some time now these equilibria had been threatened from various quarters and, in certain regions more than others, were close to the critical threshold of social instability.

The new wave of protest that broke out in that situation had an extremely composite character, in aims as well as in social components. This can be ascribed to the type of mechanism of change that had set the protest in motion. It should also be related to the type of non-linearity featured in the limit-conditions reached by the existing social system.

As we know, the movement started in 1811–12 in the midland countries of Nottingham, Derby and Leicester, and involved above all, the framework knitters and stockingers. It rapidly spread to the northwestern countries of Lancashire and Cheshire among the cotton weavers. The latter mostly employed the hand-loom: cottage industry was still very much the rule. In this sector the factory system was only in its infancy and involved, above all, the production of yarn. In weaving power looms were still very few (not more than a dozen at work in 1813)[30].

The protest now spread to the wollen producing centres of Yorkshire and mostly involved the workers skilled in cropping.

What were the impulses behind a movement affecting workers from such different areas, sectors and typologies of production? They were not, to be sure, the sole protagonists of those protests, they sought and managed to establish significant links with one another: the framework knitters and stockingers were supported by many of the master workers who employed them; the weavers in cottage industry in the cotton districts found themselves side by side with the factory workers in agitations at Stockport, Bolton, Middleton and Manchester[31]. More or less everywhere the struggles of the workers were interwoven with protests against the high cost of living, the latter involving figures from quite disparate social backgrounds and, both in towns and villages, easily assuming the features of popular agitations on a broader front.

The growth of the "movement" and its spread from one area to another, however, also affected other sections of the working population (unemployed field labourers or those reduced to hand-to-mouth conditions, other categories of poor workers whose situation had worsened in one way or another). To these were added sometimes numerous groups of people already long accustomed to leading a life of vagrancy.

Thus we see how the nucleus of the movement is surrounded by a broader outer layer, as it were, of social instability. This has often received a reductive treatment at the hands of historians, has been regarded as a sort of parasitic fringe of the Luddite movement, substantially detached from it.

But in our formulation the adoption of classificatory criteria on the basis of abstract, prejudicial discriminations can only mislead. Careful analysis, mindful of the complexity of the social dynamic, will hardly support such distinctions and exclusions. Certain aspects of the social protest may turn out to elude the scheme of interpretation or be awkward for it. But the literature on the relations between the working classes and the "dangerous classes" is already too rich for us to adopt surreptitious distinctions reproducing the fears and exorcisms of the ruling classes of that period, or the political prejudices and social discriminations with which the leaders of the workers' and labourers' movements subsequently regarded those antecedents of the social protest.

Rather, if we would understand that cycle of protest that took its name from the Luddite movement, we must explain its highly complex, ramified character—spurious, if you like, but all the more significant for that. And, indeed, the effect of that "outer layer" of the movement, as described above, seems inseparable from the "prismatic" composition of its central nucleus.

I have argued that the type of mechanism of change responsible for the passing of the critical threshold of social instability arose out of the establishment of relations between previously independent events since it altered a working regime based on the complementarity between different production activities: and that this accounts for the composite character of the movement under way, for the plurality of social demands inherent in it. The presence of various social impulses and components, the multiplicity of needs that stimulated the process, the complexity of the underlying social dynamic, should be set against the type of non-linearity featured in the previous working regime, i.e. the limit-conditions in which the pre-existing equilibria were disrupted.

Like what we have seen in the case of Italy, in the context of English society in the early nineteenth century, and in certain areas more than others, the regime of complementarity affected, above all, the contiguity between rural economy and non-agricultural production, especially in the traditional kind of textile work. When the critical point in the equilibrium of subsistence and the relation of work and system of life was passed—as a result of the mechanism of change referred to—the effects were more nearly felt in those parts of society and where the contiguity had most importance.

Now, whereas in Italy the mechanism first affected agricultural production, and textiles thereafter more as a side-effect, in the case of

England although the same mechanism was at work, by reason of the different events that had entered into relation with one another the first areas to be affected were those of textile production (import-export cycle of cotton and trading in other manufactures, especially woolens). Nevertheless, in the one case as in the other, the links between what took place in industry and what took place in agriculture remain of the essence.

As regards the previous limit-conditions, it will be useful to compare what occurred among the labouring population in the rural Po Valley before their protest began, with a similar phenomenon involving the English weavers during the closing decades of the XVIII century and the first decades of the XIX. This period witnessed a heavy demand for manpower in the weaving areas, while at the same time this manpower substantially lost status and came to depend more and more on the new employment.

However this change cannot be accounted for by innovations in production and techniques—in this case, the introduction of the power loom and the factory system. The innovations were still in embryo as far as their application was concerned; the new factories dealt, if at all, with the first stages of production—mainly the production of yarns.

The number of weavers was increasing in the traditional sectors, where the hand loom still dominated, whether in cottage industry or manufacturing. What actually happened was that for many landworkers, either wage-earners or smallholders, a job that had once served mainly to supplement the family income—performed either by the head of the family or entrusted to other members, according to a traditional division of work by alternation of tasks and regulation of roles—was by now almost the only gainful activity they could depend on.

But the erosion of the margins of subsistence and the crisis in the rural community ultimately affected others, too: not a few of these were migrant artisans who went to swell the ranks of "journeymen weavers". And these latter continued to be fed from the pocket of endemic unemployment in the north of England. In this way a mass of journeymen weavers was growing all the time; a work force ever more dequalified and underpaid as compared to the traditional modes of employment and ancient regulations of the trade.

With the labor market already saturated, this workforce, partly superfluous, semi-employed, ready to accept wage cuts in the internecine competition within its own ranks, was completely at the mercy of every oscillation in the market and any imposition by the employers bent on exploiting the situation.

In practice, what happened was that many merchant-entrepreneurs "speculated" both on the trend of the market and on the employment of labor conjointly: when the market contracted, they gave the work out to outworkers who were prepared, at that time, to accept any pay rate; when the recession finished they flooded the markets with goods produced at low prices, thus continuing to keep wages down.

This situation had already reached its limit. And now the extension of its fluctuations owing to the combination of different events, as we have seen, led rapidly to the passing of the critical threshold of social instability.

Before all else, there was an increase in the serious tensions that affected the labour market in the weaving sector. The waves of emigration became more intense, especially from Ireland. The effects of the exodus from the rural areas began to multiply.

As against that, the repercussions of the simultaneous crisis in agriculture put paid to the remaining chances of eking out a subsistence by supplementary field labour. The weavers, who more than any other workers had maintained contact with the rural economy, were especially hard hit.

In similar fashion to the one we have described in the case of Italy, the English textile entrepreneurs reacted to the crisis by adopting a predominantly conjunctural logic. Restructuring of the production cycle and work systems was not complete but only partial, according to the dictates of opportunity and calculations of what would be of immediate convenience.

The upshot was that new elements in the modes of employment and exploitation of the work force were, indeed, introduced; but far from replacing the old ones, they were rather added to them or grafted onto them, in such a way as to render the condition of the workers intolerable and full of unacceptable contradictions.

I have already underlined how this actually constituted a second mechanism of change, added to the first and combining with it to lead to a passing of the critical threshold of instability. And it is highly significant that the same phenomenon, owing to the same mechanisms, should also be found in the case of the Italian textile workers.

In their attempt to occupy the diminishing areas of the market, the larger entrepreneurs ruthlessly sought to exploit the newly available manpower, dequalified and lowly paid (what the old workers called "illegal" because it was employed in defiance of the established regulations and customs). At the same time, by producing lower quality goods at reduced prices, they tried to avail themselves of partial mechanization in the production cycle, above all in its first stage, for the production of half-finished goods. This mechanization was already under way (for example in yarn spinning) or was available and easy to adopt (as in the case of large warp looms). In the second stage of production, that of "dressing", certain processes could also be simplified with mechanical devices that had already been long available. It was only a matter of overcoming the workers' resistance, backed by ancient statutes.

And in the new situation of general crisis, with a weakened and de-qualified work force, the age-old resistance would be all the easier to break.

But that must happen, or so it was claimed, without disturbing the old system of working relations (now preserved only to the entrepreneur's advantage) as they regulated recruitment, intermediation, tendering of work, piecework. This system was not abandoned but rather exploited in the new situation by being adapted according to convenience.

Thus it came about that the middlemen travelled the country areas seeking to place large quantities of yarn with individual weavers, who were always on the lookout for work and compelled to accept cuthroat conditions by the middlemen. The weavers were swindled over the weight of the half-finished article, for which weight they were, however, responsible on delivery of the goods. Carriage of goods before and after working was no longer undertaken by the weaver, yet he had to defray the expense. Workers whose situation did not permit purchase of a loom were gathered together in "factories" with hand looms, where the conditions were even worse. Nor was the move to "factory" work any sort of guarantee for the worker: when it suited the merchant-entrepreneur to cut production, this category was the first to be dismissed. More generally speaking, the hiring of looms, which was customary among the master-weavers who recruited journeymen on the spot, was now used by the new middlemen as a further device to keep wages low.

Overall, the system of cottage industry was overturned: the journey-man weaver lost any kind of independence but had to meet obligations and was paid just as though he still had the faculty of regulating his own work.

The worsening of working conditions—due to new elements superimposed on the old systems of employment and exploitation of manpower—also hit the most skilled workers who were employed in the finishing trades. Indeed, the contradictions in their case were even more glaring. What did it mean for these workers to have to fall back on lower quality products at cheaper prices? How were they affected by the merely partial reorganization of the production cycle? The stockingers and framework-knitters in the manufacturing districts of Nottingham,

Leicester and Derby were now required by the merchant-entrepreneurs to supply large pieces of material woven on large looms, which would then be cut up into the shape desired and sewn together side to side ("cut-ups"). The finished article was of inferior qualify but could be mistaken for the one obtained by traditional methods of finishing. Yet the workers' skill and rates of pay continued to be estimated according to the old logic of piecework, i.e. based on the quality of finish of the article. Thus he ended by being paid as for a coarser article, for which he was nevertheless judged responsible. Not only that; the system of fines for defective or careless work, which was no longer the worker's fault, became arbitrary and vexatious. Here again, under the old piecework system and while the worker still had complete responsibility in "dressing", it was understandable that he should compensate for any shortcoming on his part. Under the new conditions the responsibility had been taken away from him, yet the costs of operations regulated by others were still subtracted from his pay. This was equivalent to real wage cutting. Nor, though the wage-cuts came in a time of worsening conditions, was this all: the growing dequalification of the work force was such that it could now be exchanged or substituted with common laborers or children.

Similar processes occurred among other craft workers. In the wool manufacturing districts of Yorkshire the croppers and combers were also paid on piecework. The first cropper's pay, for example, could vary by as much as 20% per piece according to the quality of the work. They, too, were responsible for the entire dressing process. But the introduction of quite simple machinery, already long available, such as the gig-mill and the shearing frame meant that those steps that had demanded greatest skill on the part of the worker could now be skipped. So we can understand why skilled workers objected to these innovations. But their hostility and protest become all the more comprehensible when we bear in mind that these partial innovations had been superimposed on a system that nevertheless retained the old pattern of piecework with regard to pay and regulation of work.

Here we find not merely threatened loss of status, as many historians have, quite correctly, emphasized[32]. The process led, rather, to conditions of work and pay that were as contradictory as they were intolerable. And it was this specific dynamic that ultimately led to conflict—as we have already seen in the case of Italy. In brief, we find a second mechanism of change that—in the case of the English weavers as in that of the Italian textile workers—is added to the previous mechanism and combines with it, so that the limit-equilibria of social instability are transcended.

At which point protest was inevitable and in the nature of things. And, indeed, in the case of the Luddite movement the protest also assumed a typical trend and connotations that can be referred to the type of non-linearity featured in the previous limit-conditions.

The movement began in Nottinghamshire in March-April 1811. Framework knitters and stockingers destroyed about 200 looms in opposition to the new system of working on cut-ups. November saw the revolt spreading to Derbyshire and Leicestershire. In north-west Nottinghamshire in the first months of 1812 looms were broken at a rate of 200 per month.

As has been stressed, this type of action was anything but riotous or improvised; on the contrary, it demanded iron discipline and precise, well-led organization. It also required substantial support from the local population[33].

The aims were very precise, as can be seen from the proposal for a Bill "Preventing Frauds and Abuses in the Frame-work Knitting Manufacture" that the "Nottingham Committee" intended presenting to Parliament. The proposed bill called for a mark to distinguish quality products from shoddy ones, the prohibition of inferior copies of high quality articles, compulsory use of the "rack" in fixing pay and a price list to be displayed in every workshop; and lastly regulations concerning the hire of looms[34]. By the end of April this petition had gathered 10,000 signatures. The list shows how the framework knitters and stockingers did not lack support from significant sections of the population, including the rural ones[35].

In the preceding months other petitions, this time from the weavers to obtain legislative redress against unscrupulous entrepreneurs, travelled over an extraordinarily wide area, picking up 40,000 signatures in Manchester, 30,000 in Scotland and 7,000 on Bolton[36]. Here was the most striking proof of how the protest movement could expand still further, as indeed it did, spreading to the north-western countries.

In particular, in the cotton manufacturing districts of Lancashire attacks on factories using mechanical looms and specific agitation by weavers for a minimun wage were closely combined with strong and frequent protest at the high cost of living. On March 20, 1812, at Stockport the warehouse of Ratcliffe, an early employer of the power loom, was attacked. In the weeks following a crop of serious riots broke out against the high cost of living, in Stockport, Manchester, Oldham, Ashton-under-Lyne, Rochdale and Macclesfield[37]. On April 20 at Middleton another factory employing power looms was attacked. Simultaneously the food riots spread everywhere: in the north from Barnsley and Leeds to

Carlisle; in the south at Bristol. In Cornwall the miners came out on strike and demonstrated at Plymouth and Falmouth. While at Sheffield the arms magazine of the militia was looted. All these demonstrations had a high political tone: in many cases the aim of the demonstrators was to enforce a "popular maximum"[38].

When on May 11 that year the Prime Minister, Perceval, was assassinated, in Bolton, Nottingham and London the event was greeted with open jubilation.

It was anyway only logical to expect that on reaching its climax the protest should assume political features. This was already evident in March when at Manchester the wrath of the weavers was turned against a loyalist demonstration by the merchants of that town in favor of the Prince Regent[39].

In the meanwhile the movement had also spread to the woolen manufacturing districts of Yorkshire. In January a gig-mill working in Leeds was attacked. In February the attacks spread to Huddersfield and the Spen Valley, where the gig-mill and the shearing frame were most in use. The woolen workers of the West Riding threateningly proclaimed their close alliance with the workers in the cotton districts of Manchester, Oldham and Rochdale, as well as with the Scottish weavers and even with the Irish "papists"[40].

The climax of the Yorkshire movement came in April 1812 just as the struggle in Lancashire was growing fiercer. In the case of the woolen workers, too, hostility was directed not soley against the entrepreneurs but was also aimed at the government and the Crown itself. In April-May 1812 the unrest reached a peak at its various epicentres; but at the same time the movement presented such extensive ramifications and implications that one may justly speak of an "insurrectionary fury" involving the entire country[41].

In the West Riding, too, as already in the Midlands, the widespread feelings of public opinion were closely connected with motives behind the unrest[42]. Yorkshire, in any case, did not differ from other regions, in as much as the main component of the movement, in this case represented by the woolen workers, was not an isolated one: the protest mobilized several other categories of workers. Among the prisoners sent for trial before a Special Commission at York there figured a woolen spinner and 3 cotton spinners, 2 clothiers, a carpet weaver, 8 laborers, 3 coal miners, a stonemason, a carpenter, 3 shoemakers, 2 tailors, a hatter, a hawker, and a waterman[43].

At that moment some 12,000 troops were employed in the suppression of the protests and riots up and down the country; as has been rightly pointed out, this was a larger force than Wellington had at his disposal in the Peninsular War[44].

By now it had become clear that police measures alone would not suffice; action would have to be taken at political level. And the Government did indeed try to allay the tension, while at the same time reinforcing the solidarity of the owners, by repealing the "Orders in Council". The owners, on their side, called a momentary halt in that partial, one-sided restructuring of the work system and production that had compelled the workers to protest.

Yet even when the suppression was successful, the fires were not quite put out. In many cases men turned to collecting arms and gathering funds together. There were further, more sporadic incidents in 1813–14 and 1817[45].

With its epicentres and the motives initially behind it, that remained its most typical features, the Luddite movement in its climactic stage can be seen as a sort of nucleus of a vaster movement that soon took on other qualities and extended its roots into different areas and strata of social reality; it was also able to exert ever ampler and more violent repercussions on the country's political life.

If we look at the broader social spectrum of the movement, at the deep-lying motives and irreversible features of its dynamic, we shall understand that although it represented an opening stage in a new type of social protest, yet this stage, far from burning itself out, was destined to reappear in other forms in the period following.

Viewing it in this way, we cannot ignore the various elements that were affected by the struggle and the different layers of society it involved. By the same token, we must take account of the plurality of levels at which it was undertaken, not excluding the legal and the political; those latter were sometimes involved hand-in-hand with violent action, sometimes as an alternative to struggle—as was the case with the protest of the weavers of Glasgow and Carlisle aimed at obtaining regulation of wages and apprenticeship[46].

Nor can we underestimate the extent and significance of the political elements in the much discussed events in Lancashire.

Organization reached a high level and was not limited to single typologies. Not merely do we find the undercover organization of the weavers, operating in at least twenty places large and small in Lancashire, with secret committees reaching out to involve a great variety of trades[47]. Due weight should also be given to the role played by organizations like the National Weavers' Union, which was active from 1809 to 1812 despite prohibitive laws, or by horizontal organizations representing various categories, like the "Committee of Trades" operating in the Manchester—Stockport district.

Any analysis of these events that goes beyond the merely superficial cannot miss the elements destined to recur, in less fugitive, less embryonic guise, in the later stages of the social conflict[48]. Above all, if we take proper account of the social dynamic that underpinned what happened, we shall be in a better position to set it in the framework of a larger process of crisis and transformation of which our narrative has shown only the beginnings.

And at this point it seems to me that comparison between the contexts and periods, however different, which saw the eruption of a new social movement as I have here described it—in the case of Italy as in that of England—will confirm the hypotheses I put forward.

Obviously an analysis conducted in terms of examples has its limitations. But it is clear enough that by referring to the mechanisms of change able (in the presence of very extensive fluctuations) to lead to the passing of the critical threshold of social instability, we can investigate the motives and ways of the rupture represented by the new dynamic of protest as it appeared in the two contexts. The social movements in Italy and England showed a composite character, were full of contradictory features that elude the traditional schemes of interpretation, and this reflects the complexity and non-linearity of the social dynamic that accompanies them and explains them.

Above all, I think it has become sufficiently clear that one can perform a comparative analysis without sacrificing any of the specificity of trends and motives in the social movement in two such different contexts. Indeed, it is by investigating the features peculiar to each that, in a morphological analysis of social change, we can make comparisons and verify how far the results correspond to our expectations.

I also intended to evidentiate the non-deterministic, non-random character of the morphological change that arises beyond the critical threshold of instability. The trends in the process of crisis and transformation I have analyzed would seem to confirm this important connotation.

A SYNCHRONIC COMPARISON: EUROPEAN EMIGRATION IN THE LATE 19TH AND EARLY 20TH CENTURIES.

It is now time to test the validity of our formulation in another dimension of analysis.

As I said above, when we attempt to explain the different outcomes of a single phase of crisis and transformation, we shall find it useful to refer to mechanisms of change that, in presence of abnormally wide fluctuations, are capable of taking the process beyond the critical threshold of social instability and thus setting in motion a new dynamic. Thus we see how, when the situations of instability vary, a single mechanism of change—not a mechanism of the same type, but that mechanism itself bearing on the same events —can start different social dynamics.

Examining the conditions of Italian society at the close of the 19th century, we find that in those same years and same country the morphology of change turned out to be quite different. The same mechanism of change we identified at the root of the social protest in the rural Po Valley and surrounding areas, where there was greatest complementarity between agricultural and non-agricultural production, led to a completely different outcome in the rural areas of Southern Italy.

Here we have the same relations between events previously independent of each other: the growth of new and vast cereal-producing areas in the Americas, north and south; the new importance of rice cultivation and textile plants in South East Asia; the new systems of ocean transport by steam vessel; the cutting of the Suez Canal; the massive expansion of railway networks in many European countries. But that mechanism of change set in motion a quite different dynamic in the South of Italy, characterized this time not by a social protest but by mass emigration.

The mechanism of change led similarly to the passing of the critical threshold of social instability, but, acting in different conditions, it led to a different outcome. This was still in relation to the type of non-linearity featured in the previous working regime of the system.

The country areas of South Italy were still divided up predominantly into large estates. A semi-feudalism persisted and the estates laboured under the typical burden of rentier-owned property; the rentiers being large-scale, but also medium and medium-small. In the extensive cropping of the large estate, as in the afforested areas (mostly along the coasts), the workforce was employed in very reduced, periodic labour, limited to seasonal jobs, that could not be supplemented with other periodic activities in other sectors of agriculture.

Alongside the labourers there were other workers operating in not very dissimilar conditions: *metatieri* and *terraticanti* on the large estates in Sicily, and similar figures in the hinterlands of mainland Southern Italy. Their engagement was by contracts that, formally speaking, stipulated co-participation, with rent or partial tenancy; in actual fact, the one-year duration and the pre-payment for seeds and wear and tear of implements, meant that the workforce was compelled to operate in highly precarious conditions and closely depended on its employers. All these workers had no other possibility of supplementing their income and lacked any security or broader basis of subsistence; and this, together with their absolute social dependence, deprived them of the slightest chance to protest or bargain.

Here, too, we find a limit-condition of social instability, but of a qualitatively different type, by reason of the particular fluctuations in the labor market and in the margins of subsistence. The difference lay in the fact that the workers' link with the land was much more aleatoric, whether regarding seasonal movements or annual contracts; and, as we have seen, they lacked the possibilities, available in central north Italy, of supplementing their incomes with other kinds of labor.

In the North the land was split up into much smaller properties with laborers and owners permanently present on the land. This made for a much more elaborate and significant tissue of society. The stratification of the population, the social relations and the customs of the rural community had a consistency quite different from those of the South. And all of this went to make up a wider basis for the day-to-day economy, for the margins of subsistence and the very modes of living of the working population.

The existence of the southern peasants rested on no such basis. The tissue of their economy and society had, as it were, a much coarser mesh. The rural population was gathered into large townships, often at considerable distance from the lands they were periodically sent to till. This resulted, perforce, in a weaker, more uncertain relation between them and the land.

Which is to say, conditions were more precarious, not only as regarded working relationships but also with respect to the entire social reality and modes of life, as a whole.

In this situation of general weakness and instability, the peasant suffered all the more from his absolute dependence on the landowner or, worse and more frequently, in his relationship with the various sub-tenants; with the landowners' tentacles finally reaching him in the form of the tax-gatherer and the field overseer. Instability often attained the critical point of permanent indebtedness, which could be more and more difficult to sustain owing to the practice of usury. So that the rupture of that relationship—which would mean the possibility of unemployment—was always imminent. A very small thing would suffice—a poor harvest or bad weather, an adverse decision by the tax-gatherer or the field overseer—and the labourer might find himself unable to get work or *de facto* expelled from the land.

For there was not the minimum economic or social basis on which resistance or protest could be organized.

Here we see the type of non-linearity in the working regime of the existing system that explains how, when the limit-conditions of instability reached their worst, emigration came to be seen as the solution.

An analysis performed at this level can, indeed, explain why, in the same moment in the process of crisis and transformation, in the same country and in presence of the same events, such different phenomena could occur: on the one hand, mass protest; on the other, large scale emigration.

In practice, the difficulties posed by a problem of this type have severely tried the traditional schemes of interpretation. It has frequently been found that economic "causes" will not suffice to explain these differences in social behaviour. At which point subjective factors have been adduced, of the vaguest, most general kind: differences of mentality, more passive attitude, inability to react and protest.

On the contrary, the formulation I propose deals with the mechanism of change due to the establishment of relations between events that were formerly independent of one another and this makes it possible to compare, on synchronic basis, the various trends of the social dynamic; which enables us, on the one hand, to account for differing outcomes such as the social protest and the mass emigration in Italy at the end of 19th century and the start of the 20th century, and, on the other, to compare similar outcomes in different contexts and countries. The emigration from Italy can thus be set in the broader framework of the large-scale emigration from southern and eastern Europe in the same period.

Obviously the new relations established between the events we have mentioned could not help also having repercussions on the flow of emigration that had already begun, i.e. from Germany, the British Isles, and the Scandinavian countries. Here, too, we see a considerable increase as compared to trends prior to the 1880s.

But it was the dynamic of emigration, set in motion by the same motives and in the same ways, as we have seen in the case of Italy, that provoked a new mass exodus: new, that is, by reason of its scale, which outdid anything that had gone before; and new in so far as it involved vast areas of eastern and southern Europe that had hitherto played little or no part in the overseas emigration.

The result was, then, two great waves of migration from Europe across the Atlantic, in rapid succession.

The first occurred in the 1880s, with migration from new areas added to increased migration from the old ones (Germany, Ireland, Scandinavia). As compared to previous averages, German migrants doubled in the period 1881–85, reaching 857,000, only to lessen in the years following. The exodus from Britain also doubled, with about 1,200,000 people leaving the country in the first and second half of the 1880s, and emigration from Britain was to remain high. Scandinavian emigration also grew sharply, only to drop in the period following.[49]

Simultaneously we find the first high levels of emigration—the new type of emigration—from Italy, Spain, Portugal, Austria—Hungary and Russia. In Italy the total for the period 1886–90 reached 655,000 people. And this growth largely involved transatlantic emigration. In these first two decades Italy sent even more people to the Americas, with an increase from 80,000 in 1880 to 240,000 in 1900.[50]

The second and even stronger wave came in the period 1900 to 1914. New motives for emigration were conspicuously at work; in the space of a few years exodus from the rural areas had reached startling figures.

In this second phase Italian emigration, which had been growing steadily (810,000 in the period 1896–1900), attained a total of 1,970,000 in 1906–10; nor was it to slacken by much in the five years following (1,546,000).[51]

While emigration from Spain and Portugal was not so large in absolute terms, relatively speaking it followed a quite similar pattern of increase, with Spain reaching a maximum of 318,000 in 1906–10 and Portugal a maximum of 217,000 in the five-year period following.[52]

As for the Austro-Hungarian Empire, the first decade of the 20th century witnessed a massive exodus: about 1 million persons left Austria-Hungary in the first half and 1,327,000 in the second[53].

Transatlantic emigration from Russia became equally substantial starting as far back as the eighteen-nineties. And between 1900 and 1914 the exodus, bound mainly for the United States and to a lesser extent for Canada, Brazil and Argentina, involved 2,500,000 people.[54]

If we break down these national statistics and examine the regional situations and social structures in the areas of European agriculture most affected by the phenomenon of migration, the comparisons will appear highly significant.

The regions in question accounted for a large portion of grain production in Europe; but at the same time they were peripheral areas, the most backward among the rural zones of central-eastern and southern Europe. It was not merely that they felt the economic-commercial repercussions of the crisis; they had also reached other limit-conditions of a dynamic, more properly social in character, that were affected by the mechanism of change.

The analysis given above, with reference to the limit-conditions of social reality in the country areas of southern Italy can, *mutatis mutantis*, also be applied to other contexts.

Portugal offered a similar difference between the rural parts of the North and those of the South. In the southern plains of the Algarve and Alentejo, agriculture was dominated by the *herdades:* these were large estates, to the number of a few thousands, covering almost the entire region. The day-laborers, whose tiny allotments nowhere near guaranteed a subsistence, were concentrated in the villages.[55]

In Southern Spain, too, large estates of more than 100 hectares accounted for 55% of the agricultural land; whereas less than one quarter was split up into tiny lots worked by 97% of all owners.[56]

In the rural areas of central-eastern Europe, during the second half of the nineteenth century the "Prussian model" of agrarian development took root. The reforms of feudal relationships that had succeeded one another in the various contries in that part of Europe between the 1820s and 1860s had scarcely affected the large estates held by the nobility. In one way or another, these had been enlarged, at the peasant's expense, by the addition of common lands, and a general change came about in customs and traditional relationships. Large masses of "freed" peasants were without land. While there was a slow transition from regulated labor to wage labor, very often the two were combined, greatly to the disadvantage of the laborer. The phenomenon varied considerably from region to region in the Austro-Hungarian empire, the Russian empire and Romania.

In Romania and in 17 out of the 47 provinces of European Russia the systems of payment by labor (*otrabotka*) persisted. The peasants worked the land with their own implements and draught animals, where they possessed any; otherwise they received loans which had to be repaid or renegotiated when the produce was distributed, an improper, unstable sort of "share-cropping". Also the laborers employed in the mass seasonal tasks of hoeing and reaping had to repay the owner's advance subsidies.[57]

Aside from all the differences of socio-historical context, the analogy between modes of employment and exploitation of manpower in central-eastern Europe and the large estates of southern Italy becomes clear. Equally serious were the extreme instability and close relation of dependence, leading to a state of perennial indebtedness.

In Hungary the system of labour service was less widespread and was sooner abandoned. Nevertheless, the unstable working conditions of the farm labourer employed on the great estates, and his dependence on his employer put him at a similar disadvantage. In 1885 the larger estates, with more than 1,000 *arpenti* (575 hectares), covered 55% of the surface of Hungary and were owned by 9,000 families. Leaving aside the minority of rich "kulaks", the indigent smallholders were forced to labour on the large estates in order to survive. Yet they made up only a portion of the reserve of agricultural manpower. To them were added 3 million landless peasants who worked as small share-croppers (*kleinpächter*), farm servants and above all, seasonal workers[58].

For this last category the relationship with the land was a completely labile one, but for the others it was also extremely weak. The smallholders themselves, whose subsistence, like the rest, depended on the possibility of employment on the great estates, found themselves in a very subordinate position. Their situation featured the general characteristics of the social hierachy and reflected the gulf that separated those who worked on the land from other layers of society. And it was this social, as well as economic, weakness that condemned these workers to emigrate—as we have already seen happening in the case of the large estates of South Italy.

As for Russia, the slow transition from regulated labor to wage labor merely reflected the backward state of those huge rural areas and the contradictions present in them. Their social structure was characterized by strong disequilibria.

Following the reform of 1861, the large estates had not substantially shrunk in extension and had managed, in certain ways, to consolidate. As compared with these, there were the small and very small peasant holdings. According to some statistics of 1877, the great estates (over 500 dessyatins = 509.25 hectares) accounted for only 5.9% of all owners but occupied 80.5% of total land. Small holdings (not more than 10 hectares) accounted for 50.8% but had shrunk to 1% of total land area.[59]

Even taking all the various forms of land tenure associated with the *nadyel*—over and above tenure in the strict sense—into account, the disproportion was still enormous. According to official statistics of 1905, in European Russia, as against 30,000 large owners controlling 70 million hectares, there were 10½ million peasants sharing the same amount of land, with an average of 7 hectares each.[60]

But, between these two layers of society stood the class of rich peasant (*kulaks*), whose importance had increased. The liberalization of the property market, the rise in land prices (which had trebled in the thirty years following the reforms), and the similar rise in rents, had all favored the expansion of this class. In 1905 they accounted for a considerable extent of Russian land, namely about one fifth.[61] In several regions they had a stronger hold over the weaker classes of peasant even than the aristocratic landowners themselves.

The situation of the poorer peasants had substantially worsened. The common land of the *mir* was threatened on all sides. Moreover, the periodic redistribution of holdings in ever smaller portions of unequal fertility, and the obligation of three-year rotation, detracted from the remedy this institution constitued for so much of the population. To all this must be added the sharp increase in the numbers of freed, but landless, serfs: in the last decades of the XIX century and up to the Great War they had swollen from 3 to more than 5 million.[62] Other elements should also be taken into account in describing so vast and complex a reality.

But a complete analysis would take us too far from the point, and besides, it could not ultimately avoid dealing with the motives and modes of the dynamic protest in the country areas that represented an essential component of the revolutions of 1905 and 1917. The broad outlines of the picture given here will serve merely as a background to assist in understanding the character of the social stratification in those regions of Russia affected by the dynamic of emigration.

Transatlantic emigration from Russia came mostly from the Baltic regions: Poland, Lithuania, Finland and, to a smaller extent, Estonia and Latvia, plus part of the Ukraine.[63] The limit-equilibria of the social setup in these regions were due to a mainly oligopolistic regime of land tenure, to the presence of landless peasants and farmers working estates so tiny as to be hardly worthy of the name. To sum up, other components of "equlibrium" in the pyramid of land tenure were lacking, as were traditions and institutions that were more typical and more permanent in the Slav rural world (*obshchina, mir, zadruga*, etc.).[64] From this point of view the Polish country areas constituted an exception.

The landless peasants, like the tiny smallholders (who might more properly be considered as a kind of agricultural sub-proletariat), made up a "total" reserve of manpower for labor on the great estates, with no other possibilities of subsistence. In these conditions the time-lag and contradictions in the transition of the great estates towards more modern forms of employment of labor had repercussions on these very labourers. They were fettered by the semi-feudalism that remained without receiving any compensation in right of tenure, and were vulnerable to the uncertainties of wage labor without any possibility of bargaining. Theirs was thus a situation of maximum instability. And this, as we have seen, represented the limit-condition of social dependence and precariousness in which a dynamic of migration could the more easily assert itself.

A look at the data will show how in Russian Poland one half of the migrants were landless peasants, followed by the very humblest of the smallholders (the latter left behind them "estates" that, in two thirds of cases, covered less than 5 hectares each, and in one sixth of cases, less than 2 hectares).[65]

But in Austria-Hungary, too, the Polish regions led the emigration, followed here by the Hungarians, whose social stratification in the country areas was, as we have seen, not dissimilar.

It can hardly come as a surprise, then, that between the end of the XIX century and the start of World War I Poland as a whole sent out 3,500,000 emigrants. While the 2 million who left Hungary made up a little less than half of the transatlantic exodus from Austria-Hungary[66].

I think this picture offers sufficient evidence of the type of non-linearity featured in the structures of rural society in those regions accounting for the largest emigration across the Atlantic at the close of the XIX century and the opening of the XX.

In spite of all the considerable historical discrepancies, these characteristics can be summed up in the following way. The countryside of southern Europe like that of central-eastern Europe, was made up of extensive agricultural areas mainly devoted to cereal crops and retaining traditional crop-rotations and methods of working. The most significant, most often recurring feature of those social structures was the regime of tenure, which was characterized by the clear prevalence of large and very large estates either originating among the aristocracy or taken over by them. During the XIX century, even during the second half, the aristocracy had generally succeeded in maintaining its position on the land; in several cases it had actually enlarged its estates at the expense of peasant property and manorial or common lands, as well as civic custom, etc.

Bourgeois landowners, of varying origin and distribution, had certainly made considerable progress, but accounted for proportionately less of the property or had less importance, in the regions we have examined, than in other areas of the same countries; whereas there was a greater persistence of semi-feudalism.

The pyramid structure of social relationships in the country areas was thus a very lopsided one. At its base, the two broadest stratifications were made up of tiny "particle-like" smallholdings and landless peasants.

The first of these must be distinguished from the small independent property owners however poor the latter, however close to the limit of subsistence. The independent smallholding was typical of certain agricultural areas (mountain regions, above all) or others in which it made up the traditional social fabric. This was the case, for example, of alpine valleys in the Veneto and Austria, in north-central Spain or in the rural areas of south west Germany. And these wretched smallholdings did not escape the effects of the crisis at the turn of the century, playing their part in the emigration.

But the smallholding we have described as "particle-like" was a different matter: and not merely because it was even tinier than the former and could not furnish anything like a livelihood for the peasant and his family. As compared to the large estate it was in the nature of an "appendix" or a sort of conterminous "fringe". The agricultural areas where these smallholdings lay were the same and the same crops were grown. Production was all the more insufficient for this reason.

The survival of the "particle-like" smallholder depended substantially on his chance of working on the large estates. He was essentially an agricultural semi-proletarian.

The "appendix-like" (or residual) and highly dependent character of his condition was in any case closely bound up with the historical evolution of that regime of land tenure. According to the particular context, this miserable smallholding was what was left to the peasant when he had ransomed himself from his feudal bonds by cession of land; or it might represent the residue of a once larger holding now eaten away by debts, mortgages and iniquitous taxation; or again, a remaining moiety after common lands had been expropriated, civic customs abolished and the traditional institutions of the community seriously weakened. It is thus only to be expected that the peasant's original dependence on the landlord should be clearly reflected in the residue of semi-feudalism and the vexatious ties involved in his employment by the great landowners.

The modes of this employment differed little, if at all, from those involving the other large stratum of land workers consisting of landless peasants: whether these were the agricultural sub-proletarians of the Slav rural areas, the Portuguese day-laborers, the co-participatory workers on the great estates in Italy or Hungary, or others. The fact, moreover, that workers of this type could also cultivate for their own ends a tiny patch of land on the outskirts of township or village, without thereby deriving an appreciable livelihood in terms of food, made them smallholders only in the most summary and hypocritical classifications. In reality, as we have seen in the case of southern Italy and similar contexts, there easily arose a whole chain of advances against wages and for hire of tools, iniquitous or annoying conditions in sharing out produce or calculating pay, further loans, more obligations, so that the situation of these workers actually became one of permanent indebtedness and total dependence. Where it might differ from that of the tiny smallholder was that the latter was more likely to possess draught animals and other implements, which could somewhat alleviate the iron conditions of his engagement.

To sum up, the semi-proletarian peasants and tiny smallholders, the unstable landworkers who took a share of the produce, the day-laborers with or without their plots of land, the landless ex-serfs either earning a wage or still doing regulated labor—all these, taken together and "structurally" considered, constituted the vast mass of the reserve labour force on the great estates. Their condition was, at one and the same time, highly precarious and strongly dependent; and this was the basic datum that characterized the critical values, already close to the threshold of instability, that recurred in that type of social structure.

As we have seen, these workers had to bear the dual burden of a slow, contradictory transition towards more modern forms of employment and exploitation of the labor force. They had lost all their rights on the land but as yet had no possibility of bargaining for the hire of their labor.

As the critical values of the social equilibrium moved towards their limits, they had no chance of "resisting", either passively or actively. In their condition we can easily identify the type of non-linearity that showed itself, in practice, more sensitive to the mechanism of change capable of setting in motion a dynamic of migration on a vast scale.

The "particle-like" peasants and the landless peasants all worked on the big estates in a range of different conditions that ran from semi-proletarian to sub-proletarian and where it was often hard to distinguish between these; The next in line on this inclined plane were the casual workers, pure and simple, who found jobs only when the great seasonal tasks were in hand and who represented the extreme point of instability, on the borderline of vagabondage. These two fundamental strata of landworkers, who in a oligarchic regime of land tenure made up the base of a highly unbalanced social structure, also represented the large mass of the population in the rural communities of those areas we have mentioned. In the large townships dotted about the large estates in the Mediterranean region, as in the small villages in the countryside of the Slav nations and the Baltic areas, they were the main, most homogeneous element in the population. The crisis that involved them had a sideways repercussion within those communities, and ultimately affected other groups, including non-agricultural workers, small artisan and country tradesmen.

So that the fundamental equilibria of the entire rural community were shaken out of true and weakened. Nor could the economic aspects of the crisis be isolated from the others. Above all, in a still highly traditional social network, the margins of subsistence, and thus the values of the "day-to-day economy", were closely linked to the modes of life, customs, models of behavior.

At the same time we must remember that we are dealing with a very coarsely meshed tissue of society, held together by few links or strong points and thus very feeble and inflexible and all the more likely to come apart. In a social reality of this kind, any large change in the limit-equilibria must have had a disruptive affect. And as it turned out, when the dynamic of migration got under way its results were to multiply as one migration led to another. The outcome can be seen in the way these communities were weakened and drained of their force, often irreversibly.

In sum, it was in consequence of an identical type of non-linearity of the social structure—the type I have described in the rural society of southern Italy and in other contexts affected by the same phenomenon of amplified fluctuations—that a mass emigration became possible, one able simultaneously to involve regions and populations so geographically and historically distant from one another.

But this dynamic should not be seen solely in relation to the motives behind the emigration: the mechanism of change that we have considered led to a completion of the circle. The same ships that, in increasing numbers arrived in the ports of Europe to unload America cereals, made their homeward journey carrying the migrant workers who were attracted to the new agricultural areas. These new waves combined to substitute the *old immigration* with the *new immigration* in the American labour market; and there was also a large movement towards the semi-capitalist economies of South America.

At this point, if we examine in what sectors of the labour market and in what conditions the immigrants found employment, we shall see how the closing of the circle involved not only the technical-economic aspects of the new dynamic but also the profound social aspects. A very precise relation was established between the limit-conditions in which the work force found itself in the regions and in the social realities from which it sprang and the modes of its employment and exploitation when it arrived at the new destination.

The highest wage-rates these workers could achieve, in order to send money back to their families or return to their country of origin with something in hand, were attainable only if they subjected themselves to the hardest conditions their capacity for self-sacrifice enabled them to survive; Their survival depended on the extent to which they were accustomed to these conditions and how ready they were to tolerate them.

The fact that they were used to existing on a bare subsistence enabled them to tighten their belts as much as possible in order to build up a small saving. In Latin America above all, the necessity to adapt to a mode of life in social-cultural conditions no less desolate and oppressive than those they had already experienced, in a completely alien setting, was the price they paid in order that the meagre sums they sent home might appreciate, thanks to the favorable rate of exchange in their native countries. Thus a second circle was closed, one more internal to the social condition of the migrants.

For them their new social condition might be said to represent a form of "resistance", however, passive, that had not been available to them in their homelands.

The relation with the type of non-linearity that characterized their original social conditions thus remained very important and active in the new dynamic. And this also accounts for the dynamic's striking ability to propagate itself, its irreversible character and, at the same time, the re-propounding of its morphology.

We have already seen how, in the case of a new type of social protest on a mass basis—an implacable, systematic movement, able to propagate itself throughout the entire system—this dynamic contributed to a more general process of crisis and transformation, in its character of vector of a new morphology of change. In this perspective we can also set the great European migration at the close of the 19th century and the beginning of the 20th.

Similar and significant phenomena can be found in the periods prior to this one. But new aspects are present. Not merely are the quantities increased out of all proportion. Nor can these quantitative data be separated from the qualitative connotations of the phenomenon. Its irreversible character, the motives and modes of its extent and duration, right up to the outbreak of World War I, were indicative of a new dynamic. From this point of view, the most significant, most striking aspects lay in the fact that a veritable dynamic circle was established. This circle consisted of the complex of technical, economic, demographic, social and cultural relations between the areas from which the exodus took place and those that attracted the migrants.

And, in effect, this represented a new morphology of change and social transformation, which was destined to repeat itself and to characterize other phenomena of large-scale migration in subsequent periods, right up to recent times.

Thus the case of emigration, like that of social protest, involved new dynamics of change that participated in a larger process of crisis and transformation: a process that saw the significant and essential features of a new and more complex social organization gradually take shape.

NOTES TO CHAPTER 6

1. This critical revision has been under way at least since the nineteen-seventies. Among the many contributions we may note that of Hobsbawm (1974), the congress *Zwischen Sozialgeschichte und Legitimations-wissenschaft* of the same year, the essay by Haupt (1978) and Hobsbawm's most recent remarks (1984), chap. I.

2. Formulations of this kind are especially common in Italian history-writing of the 'seventies: see the study by Merli (1972) and the review "Classe". The importance of "objective" factors in the technical-production process has been reaffirmed, also from other standpoints. In the area of German historiography, the book by Conze (1957) has been very influential in accrediting a model of integration of the working class: see, for example, Fisxher (1972), Schomerus (1977); compare also the various articles in the book edited by Conze and Engelhart (1979). Tenfelde, instead, sees the transformations in technique and production acting as factors of coercion towards adaptation. In the view of others, these transformations were, rather, a reaction to the shortage of labour and the demands of the workers. Yet the workers' opposition was not really capable of combating capitalist rationalization. Cf. Kocka (1969)

3. On the debate over the relation between history of the labour movement and social history, see (in the rich literature on the subject) Hobsbawm's study (1973), the contributions by Thompson, Perrot, Tilly and others to the seminar *Making and Change of Plebeian and Proletarian Consciousness, 18th–20th Century*, held at the University of Konstanz in 1977, and the papers given by Haupt, Kocka, Trempé and Salvati at the seminar on *Social History and History of the Labour Movement*, held at the Fondazione Basso, Roma, 1978. See further the studies by these and other authors where the various aspects and problems concerning the relation between industrialization and making of the working-class are examined. We quote, for example, those of Thompson (1963) and (1967), Perrot (1974) and (1978), Hobsbawm (1964), Tilly (1975), Stedman Jones (1974), Trempé (1971), Lequin (1977), Samuel (1977), Merriman (1979).

4. On this point see the critical remarks of Salvati (1980), especially pp. 6–9.

5. Cf. Masulli (1980), pp. 11–13 and 56–58.

6. *Atti della Giunta per l' Inchiesta agraria e sulle condizioni delle classi agricole,* vol. II, fasc. I (1881), pp. 225–236; *Ufficio del Lavoro della Società Umanitaria* (1904), p. XXX *et seq.*; Medici-Orlando (1952), pp. 73 *et seq.*

7. The figure of speech recalls the instability known as "Bénard's instability". A vertical temperature gradient is set up in a horizontal liquid layer. If the gradient passes beyond a threshold value, the state in which the heat is transmitted simply by conduction becomes unstable. At this point we have a phenomenon of convection, increasing the transmission of heat: the molecules move coherently, forming convection "cells" of characteristic size. In effect, Bénard's cells represent an example of how instability

can give rise to a phenomenon of spontaneous self-organization: cf. Prigogine-Stengers, 1979, Eng. ed. 1985, pp. 142–44. On this point see also Laszlo (1986), p. 36. More generally, on the interest the study of dissipative structures may assume, also for historians, see Artigiani (1987).

8. On the "la boje" movement, see, among others, Various Authors, *Braccianti e contadini nella Valle Padana, 1880–1905* (1975) and Various Authors, *Rivolte e movimenti contadini nella Valle Padana di fine Ottocento* (1984).
9. Masulli (1980), pp. 75–80; Roveri (1972), pp. 23–42.
10. Della Peruta (1984), pp. 51 *et seq.*
11. Arbizzani (1984), pp. 203 *et seq;* Giacobbi (1975), pp. 68–71.
12. Masulli (1980), pp. 145–53; Barbadoro (1973), pp. 67 *et seq.*
13. On the *Federazione Nazionale Lavoratori della Terra* see Zangheri (1960). The data are reported on pp. 6–7.
14. Arbizzani (1984), pp. 214–17; Genzini (1975), pp. 107–16; Forti (1975), pp. 420–25.
15. Valenti (1912), pp. 44 and 50–52; Luzzato (1968), pp. 169–70.
16. On the non-deterministic character of technological innovation in social changes, see the remarks by Laszlo (1986) pp. 98–102.
17. See Cafagna (1961), pp. 135–61. See also the remarks by Gerschenkron (1962), chap. IV. On the more general features of the new international cycle, compare Schumpeter (1939, pp. 397–436) and Landes, (1969, new ed. 1972, pp. 239–49 and *passim*).
18. Sereni (1966), pp. 101–277; Carocci (1956), pp. 415 *et seq.*; Villari (1965).
19. Ministero di Agritolura, Industria e Commercio, *Statistica degli scioperi avvenuti nell' Industria e nell' Agricoltura durante l' anno 1900*, hereafter quoted as MAIC, *Stat. Scioperi*, (1902), pp. XII and XXV.
20. Merli (1972), pp. 495–99.
21. MAIC, *Stat. Scioperi, 1892, and 1893*, (1894); p. 29–30.
22. Merli (1972), pp. 478–81; cf. also Ramella (1975).
23. Merli (1972), p. 476; MAIC, *Stat. Scioperi, 1897*, (1899), pp. 55–84; MAIC, *Stat. Scioperi, 1898*, (1900), p. 49; MAIC, *Stat. Scioperi, 1899*, (1901), p. 49.
24. Merli (1972), p. 502.
25. In the crucial years 1896 and 1897 the female straw plaiters accounted for 50% of all strikers in industry! Cf. MAIC, *Stat. Scioperi, 1900*, (1902), pp. VI and XXX. On the struggles of these workers, see especially Capitini Maccabruni (1964). On the proportion of outwork in Italian industry in this period, see Fossati (1951), pp. 206–44 and *passim*.
26. This percentage was calculated by adding the brickmakers (3%) to the "allied" workers in the building sector: cf. MAIC, *Stat. Scioperi, 1900*, (1902), pp. XII and XXV. In order to get an overall picture of the "industrial strikes" in the period 1892–1900 it must be remembered that, apart from the textile and building sectors, the other significant percentage, 13% in the mining sector, essentially regarded the struggles of the Sicilian sulphur miners—a very special case. Mechanical workers, including the more traditional figures of craft work and small metal goods, made up only 4% of strikers. All the remainder were dispersed in a multitude of categories and craft groups, mainly belonging to the traditional "city artisan and working class"; and they were mostly occupied in the production of goods for domestic and personal use, or in services. On this point compare the data concerning strikes with those bearing on the classification of the working population in the censuses carried out by the Italian state. See also Vitali (1970).
27. Merli (1972), p. 552.

28. On the importance of catalytic reactions in the study of far-from-equilibrium systems, see Laszlo (1986)
29. Thompson, 1963, rep. 1979, p. 603 (henceforth I quote from this edition).
30. *Ibid.*, pp. 584 *et seq;* Rudé (1981), pp. 80–85.
31. *Ibid.* But compare also Darvall's reconstruction (1934), pp. 49 *et seq.*
32. The *déclassement* of the weavers has been well evidentiated by Thompson (1963), pp. 305–11. The phenomenon had previously been analyzed by the Hammonds. (1919): in particular, chaps. IV, VI, VIII. Their treatment of the subject has often been quoted. Thompson cautions against a reductive interpretation of the workers' opposition to that process: cf. pp. 593–6 and 600–3.
33. *Ibid.*, pp. 593 and 630; see also the various examples cited of popular solidarity, pp. 637 *et seq.* This aspect is also stressed by Hobsbawm (1952), pp 63–64.
34. Thompson (1963), pp. 586–7.
35. *Ibid.*
36. *Ibid.*, p. 592.
37. Darvall (1934), pp. 49–58.
38. Thompson (1963), pp. 622–3.
39. Rudé (1981), p. 85; Darvall (1934), pp. 93–5.
40. Thompson (1963), pp. 609–10.
41. *Ibid.*, especially pp. 616 and 624. Thompson's judgment on this point leaves no room for doubt: "Sheer insurrectionary fury has rarely been more widespread in English history".
42. Documents to this effect are also quoted by Aspinall (1949), pp. 57–8 and *passim.*
42. *Report of the Proceedings under Commission of Oyer and Terminer. . . . for the County of York,* of 1813, quoted by Thompson (1963), pp. 642–3. On the relation between the categories who made up the "backbone" of the Luddite movement and the more disparate strata of other workers, see pp. 642–6 and *passim.*
44. Darvall (1934), p. 1.
45. These dates refer to episodes of a more strictly Luddite character that took place in Nottinghamshire. But other significant episodes also from Lancashire should be remembered: the march of the Blanketeers (1817), the Peterloo Massacre (1819) and, again, the destruction of looms in 1826.
46. Cf. Thompson (1963), p. 592. In Nottinghamshire, while other Luddite demonstrations were happening, the political action of the United Committee of Framework Knitters also went ahead. The Committee was reorganized in 1812 as the Society for Obtaining Parlimentary Relief, and for Encouragement of Mechanics in the Improvment of Mechanism. In the years following, the political and trade union activity became gradually less clandestine, giving way to mass protests and organized public demonstrations. as well as open bargaining. *Ibid.*, pp. 585–91. On this point see also the Hammonds (1919), pp. 229–54, Darvall (1934) pp. 139–50 and 155–59, as well as the documents cited by Aspinall (1949) *passim.*
47. Thompson (1963), pp. 618–19.
48. In the two types of association previously referred to, the organization of the workers in the course of the movement was clear to see, both in the vertical as in the horizontal dimension. The elements of continuity were already present and were later on to become more apparent. As the repression, embodied in the Combination Acts, began to slacken, the years 1829–30 saw the rise of trade federations on a national scale, representing, severally, miners, the building trades, printers and cotton spinners. In 1829–34 organizations grouping workers from different sectors were constituted, such

as the National Association for the Protection of Labour, and the Grand National Consolidated Trades Union. On the long discussion among English historians as to the precedents of the trade union movement and its continuity (or lack of such), suffice it to recall Webb (1911), Cole (1948) and among more recent studies apart from those of the authors already quoted, the ones by Turner (1962), Musson (1972), Fraser (1971).

49. Armengaud (1973), p. 68. For a review of studies on migration from the European continent, including stages in the period prior to ours, the reader is refered to J. D. Gould (1979).
50. Armengaud (1973), p. 69; Reinhard-Armengaud-Dupaquier (1968), p. 400.
51. Armengaud (1973), p. 69.
52. *Ibid.*, p. 68.
53. *Ibid.*, p. 70.
54. Reinhard-Armengaud-Dupaquier (1968), P; 401; Armengaud (1973), p. 70.
55. Léon (1978), Vol IV, pp. 433–34.
56 . *Ibid.* On the social structure of agriculture in southern Spain, see Malefakis (1970), pp. 65 *et seq.*
57. Cf. Berend-Ránki (1974a), p. 43.
58. *Ibid.;* Léon (1978), Vol. IV, p. 432. On the condition of the peasants in Hungary see also Niederhauser (1965) and Berend-Ránki (1974b).
59. These data are quoted by Lyashchenko (1949), p. 462. It must be said that the sources are somewhat incomplete and approximate, especially as regards smallholdings.
60. These data, also quoted by Lyashchenko on p. 464, are taken from Lenin (1917), p. 227. In this work Lenin makes a comparison between the data of 1877 and those of 1905, concerning land ownership in European Russia. The resulting picture is a more articulated one. Also the proportions between large and small private estates partly differ: respectively, 72.2% and 1.8% in European Russia in the late XIX century (*Ibid.*, p. 224).
61. Lyashchenko (1949) p. 462 and *passim.*
62. *Ibid.*, pp. 418–19 and *passim.* Cf. also Léon (1978), Vol. IV, p. 430.
63. Sori (1981), p. 1740. This division by region leaves out of account the Jews who emigrated from Russia (41% in the period 1899–1914). Their case is a special one and goes beyond the limits of the question we are dealing with here. But with them, as with other cases, the combination of ethnic, national and political factors with economic and social factors should be underlined. The condition of the Russian Jews was aggravated by the fact that they could not own land, and by the internecine competition that, at the lower levels of social stratification, set them against the small craftsmen and poor urbanized peasants. To all of which were added the harsh pogroms in 1902–1906 in Russia and Poland.
64. Berend-Ránki (1974a), p. 42.
65. Sori (1981), p. 1741.
66. *Ibid.;* Armengaud (1973), p. 70; Reinhard-Armengaud-Dupaquier (1968), p. 401.

BIBLIOGRAPHY TO PART ONE

Abendroth, W..—Holz, H. H.—Kolfer, L. (1967) *Gesprache mit Lukàcs*, Rowohlt Verlag, Reinbek bei Hamburg.

Abraham, R.—Shaw, C. (1984) *Dynamics: The Geometry of Behavior*, Aerial Press, Santa Cruz.

Ager, D. V. (1973) *The Nature of the Stratigraphic Record*, Wiley, New York.

Allen, T. F. H.—Starr, T. (1982) *Hierarchy. Perspective; for Ecological Complexity*, University of Chicago Press, Chicago.

Apostel, L. (1968) *Materialismo dialettico e metodo scientifico*, Einaudi, Torino.

Arnold, L.—Horsthemke, W.—Lefever, R. (1978) *White and Coloured External Noise and transition Phenomena in Non Linear System*, ''Zeitschrift für Physik B'', volume XXIX, pp. 367–73.

Atkinson, R. F. (1978) *Knowledge and Explanation in History. An Introduction to the Philosophy of History*, Macmillan Press, London.

Atlan, H. (1972) *L'organisation biologique et la théorie de l'information*, Hermann, Paris.

Ayala, F. J.—Dobzhansky, Th. (1974) *Studies in the philosophy of biology*, Macmillan, London.

Badaloni, N. (1962) *Marxismo come storicismo*, Milano.

Badaloni, N. (1966) *Materialismo*, in G. PRETI (ed.), *Filosofia. Enciclopedia Feltrinelli-Fischer*, Milano.

Badaloni, N. (1970) *Scienza e filosofia in Engels e Lenin*, ''Critica marxista''—Quaderni, No.4.

Badaloni, N. (1976) *Sulla dialettica della natura di Engels e sulla attualità di di una dialettica materialistica*, ''Annali della Fondazione Giangiacomo Feltrinelli'', XVII, pp.7–65.

Bateson, G. (1979) *Mind and Nature. A Necessary Unity*, Dutton, New York.

Bendall, D. S. (ed.) (1983) *Evolution from molecules to men*, Cambridge University Press, Cambridge.

Bergson, H. (1922) *Durée et simultanéité. A propos de la théorie d'Einstein*, Librairie Félix Alcan, Paris.

von Bertalanffy, L. (1952) *The Problems of Life*, London.

von Bertalanffy, L. (1968) *General System Theory*, Braziller, New York.

Blumental, R.—Chageux, J. P.—Lefever, R. (1970) *Membrane Excitability and Dissipative Instabilities*, ''Journal of Membrane Biology'', volume 2, pp.351–74.

Bocchi, G.—Ceruti, M. (1984) *Modi di pensare postdarwiniani*, Edizioni Dedalo, Bari.

Bohm, D. (1980) *Wholeness and the implicate order*, Routkedge & Kegan Paul, London-Boston-Henley.

Bohr, N. (1958) *Atomic Physics and Human Knowledge*, Wiley, New York.

Bradley, F. H. (1883) *The Principles of Logic*, (Clarendon Press, Oxford, Corrected impression, 1928).

Bradley, F. H. (1893) *Appearance and Reality. A metaphysical Essay,* (Claredon Press, Oxford, Ninth impression, corrected, 1930).

Brown, N. O. (1959) *Life against Death,* Wesleyan University.

Bruinsma, O. H. (1977) *An Analysis of Building Behaviour of the termite,* in "Macrotermes subhyalimus", "Proceedings of the VIII Congress Iussi", Wageningen.

Buckle, H. T. (1869) *History of Civilization in England,* 2nd ed. Longmans & Co., London.

Butler, S. (1872) *Erewhon,* Grant Richards, London.

Carr, E. H. (1961) *What is History?,* Macmillan, London.

Ceruti, M. (1985) *La hybris dell'onniscienza e la sfida della complessità,* in Various Authors, *La sfida della complessità,* Feltrinelli, Milano.

Ceruti, M. (1986) *Il vincolo e la possibilità,* Feltrinelli, Milano.

Chaisson, E. (1981) *Cosmic Dawn: The Origin of Matter and Life,* Atlantic Little and Brown, Boston.

Cohen, B. I. (1962) *Les origines de la physique moderne,* Petite bibliothèque Payot, Paris.

Cramer, F. (1979) *Fundamental Complexity, a Concept in Biological Sciences and Beyond,* "Interdisciplinary Science Reviews", No.4, pp.132–139.

Cramer, F. (1984) *Death—from Microscopic to Macroscopic Disorder,* in E. FREHLAND (ed.), *Synergetics from Microscopic to Macroscopic Order,* Springer, Berlin-Heidelberg-New York.

Cramer, F. (1986) *Die Evolution frißt ihre Kinder—der Unterschied zwischen Newtonschen Bahnen un lebenden Wesen,* "Universitas", pp.1149–56.

Cramer, F.—Freist, W. (1987) *Molecular Recognition by Energy Dissipation, a New Enzymatic Principle: the Example Isoleucine-Valine,* "Accounts of Chemical Research", volume 20, No.3, pp.79–84.

Croce, B. (1917) *Teoria e storia della storiografia,* Laterza, Bari (English trans. *Theory and History of Historiografy,* George G. Harrap & Co., London, 1921).

Croce, B. (1938) *La storia come pensiero e come azione,* Laterza, Bari

Csányi, V. (1982) *General Theory of Evolution,* "Studia Biologica Hungarica", 18, Akadémiai Kiadò, Budapest.

Davies, P. (1982) *The accidental Universe,* Cambridge University Press, Cambridge.

Deneubourg, J. L. (1977) *Application de l'ordre par fluctuation à la description de certaines étapes de la construction du nid chez les termites,* "Insectes sociaux. Journal international pour l'étude des antropodes sociaux", XXIV, No.2, pp.117–30.

Denton, M. (1985) *Evolution-Theory in Crisis,* Burnett Books, London.

Di Siena, G. (1972) *Biologia, darwinismo sociale e marxismo,* "Critica marxista", Quaderni, No.6, pp. 69–138.

Dobb, M. (1946) *Studies in the Development of Capitalism,* Routledge & Kegan Paul, London.

Dobzhansky, T. (1974) *Chance and creativity in evolution,* in T. DOBZHANSKY/F J. AYALA, *Studies in the philosophy of biology,* Macmillan, London.

Dray, W. (1957) *Laws and Explanation in History,* Oxford Classical and Philosophical Monographs.

Dumouchel, P.—Dupuy, J. P. (eds) (1983) *L'auto-organisation. De la physique au politique,* Seuil, Paris.

Dupuy, J. P. (1982) *Ordres et Désordres. Enquête sur un nouveau paradigme,* Seuil, Paris.

Eigen, M.—Schuster, P. (1979) *The Hypercycle: A Principle of Natural Self-Organization,* Springer, New York.

Eldredge, N.—Gould, S. J. (1972) *Punctuated Equilibria: an Alternative to Phylogenetic Gradualism*, in T. J. M. SCHOPF (ed.), *Models in Paloebilogy*, Freeman/Cooper, San Francisco.

Eldredge, N.—Gould, S. J. (1977) *Punctuated Equilibria: the Tempo and Mode of Evolution Reconsidered*, "Paleobiology".

Engels, F. (1886) *Ludwig Feuerbach und der Ausgarg der Klassischen deutschen Philosophie*, "Die Neue Zeit", Nos, 4 and 5. (English trans. *Ludwig Feuerbach and the End of Classical German Philosophy*, in K. MARX—F. ENGELS, *Selected Works in one volume*, Lawrence and Wishart, London, 1968, 4th Printing 1977).

Engels, F. (1894) *Herrn Dühring Umwälzung der Wissenschaft*, Stuttgart. (English trans. *Herr Eugen Düring's Revolution in Science (Anti-Düring)*, International Publishers, New York, 1939, New printing 1966).

Engels, F. (1935) *Dialektik der Natur*, in K. MARX-F. ENGELS, *Gesamtansgabe*, Moskau-Leningrad, 1935. (English trans. *Dialectics of Nature*, Progress Publishers, Moscow, 1954, Reprinted 1982).

Enzensberger, H. M. (ed.) (1973) *Gespräche mit Marx und Egels*, Insel Verlag, Frankfurt am Main.

Epstein, I.—Kustin, K.—de Kepper, P.—Orban, M. (1983) *Oscillating Chemical Reactions*, "Scientific American", volume 248, No. 3, pp. 96–108.

D'Espagnat, B. (1971) *Conceptual Foundations of Quantum Mechanics*, Benjamin, Reading (Mass).

D'Espagnat, B. (1979) *A la recherche du réel*, Gauthier-Villars, Paris.

Fain, H. (1970) *Between Philosophy and History. The Resurrection of Speculative Philosophy of History within the Analytic Tradition*, Princeton University Press, Princeton (N.J.)

Feuer, L. (1974) *Einstein and the Generation of Science*, Basic Books, New York.

Feyerabend, P. K. (1978) *Against Method*, Verso Editions, London (6th impression 1986).

Feyerabend, P. K. (1981) *Problems of empiricism*, Cambridge University Press, Cambridge.

Fiorani, E. (1971) *Friedrich Engels e il materialismo dialettico*, Feltrinelli, Milano.

Freud, S. (1900) *Die Traumdeutung*, Franz Deuticke, Leipzig und Wien. (English trans.: *The Interpretation of Dreams*, in the *Standard Edition of the Complete Psychological Work of Sigismund Freud*, volume 4 and 5, The Hogerth Press, London, 1953).

Freud, S. (1901) *Zur Psychopathologie des Alltagslebens*, "Monatsschrift für Psychiatrie und Neurologie". (English trans.: *The Psychopathology of Eveyday Life*, in *Standard Edition*, volume 6, The Hogarth Press, London, 1960).

Freud, S. (1916–17) *Vorlesungen Zur Einführung in die Psychoanalyse*, Heller, Leipzig und Wien. (English trans.: *Introductory Lectures on Psycho-Analysis*, in *Standard Edition*, volume 15 and 16, The Hogart Press, London, 1961 and 1963).

von Foerster, H. (1981) *Observing Systems*, Intersystems Publications, Seaside (Cal.).

von Foerster, H. (1984) *Disorder/Order: Discovery or Invention?*, in LIVINGSTONE, P., *Disorder and Order*, Anma Libri, Stanford.

von Foerster, H.—Zopf, G. W. (eds.) (1962) *Principles of self-organisation*, Pergamon Press, New York.

Gardiner, P. (1952) *The Nature of Historical Explanation*, Clarendon Press, Oxford.

Gardiner, P. (ed.) (1959) *Theories of History*, The Free Press, New York.

Gargani, A. (1986) *Le procedure costruttive del sapere*, Papier given at Congress on *L'imaginaire de la complexité*, Paris, June 1986.

Gerratana, V. (1972) *Darwin e il marxismo*, in ID. *Ricerche di storia del marxismo*, Editori Riuniti, Roma.

Geymonat, L. (1971) *Storia del pensiero filosofico e scientifico*, vol. V, Garzanti, Milano.

Ginsburg, C. (1979) *Spie. Radici di un paradigma indiziario*, in A. GARGANI (ed.), *Crisi della ragione*, Einaudi, Torino.

Glansdorff, P.—Prigogine, I. (1971) *Thermodynamic Theory of Structure, Stability and Fluctuations*, Wiley Interscience, New York.

von Glasersfeld, E. (1979) *Cybernetics, Experience, and the Concept of Self*, in M., OZER (ed.) *A Cybernetic Approach to the Assesment of Children*, Wetsview Press, Boulder (Colo.).

von Glasersfeld, E. (1984) *An introduction to Radical Constructivism*, in P. WATZLAW-ICK (ed.), *The invented Reality*, Norton, New York.

Glucksmann, Ch. (1971) *Engels et la philosophie marxiste*, "La Nouvelle Critique", No. 46.

Goldbeter, A.—Caplan, S. R. (1976) *Oscillatory Enzymes*, "Annual Review of Biophysics and Bioengineering", volume 5, pp. 449–73.

Goldbeter, A.—Nicolis, G. (1976) *An Allosteric Model with Positive Feedback Applied to Glycolytic Oscillations*, "Progress in theoretical Biology", volume 4, pp. 65–160.

Goldbeter, A.—Segel, L. A. (1977) *Unified Mechanism for Relay and Oscillation of Cyclic AMP*, in "Dictyostelium Discoideum", "Proceedings of the National Academy of Science U.S.A.", volume 74, pp.1543–47.

Gramsci, A. (1975) *Quaderni del carcere*, Critical edition, Einaudi, Torino (First edition, 1948–51; English trans., *Prison Noteboock: Selections*, International Publishers, New York, 1971).

Grünbaum, A. (1964) *Philosophical Problems of Space and Time*, London.

Gould, S. J. (1977a) *Ever since Darwin*, Norton, New York.

Gould, S. J. (1977b) *Ontogeny and Phylogeny*, Harvard University Press, Cambridge (Mass.).

Gould, S. J. (1980) *The Panda's Thumb*, Norton, New York.

Gould, S. J. (1980b) *Darwinism and Expansion of Evolutionary theory*, "Science", 210.

Gould, S. J. (1983) *Hen's Teeth and Horse's Toes*, Norton, New York.

Gould, S. J. (1985) *What developmental processes can teach us abaut Evolution*, Paper given at X Congress of International Society of Development Biologists, Los Angeles.

Gould, S. J.—Lewontin, R. (1979) *The spandrels of San Marco and Panglossian Paradigm: a critique of the adaptationist programme*, "Proceedings of the Royal Society of London", B 205, pp. 581–98.

Habermas, J. (1973) *Legitimations probleme im Spätcapitalismus*, Suhrkamp Verlag, Frankfurt. (English trans. *Legitimation Crisis*, Beacon Press, Boston, 1975).

Haken, H. (ed.) (1978) *Synergetics*, Springer, Berlin-Heidelberg-New York.

Haken, H. (ed.) (1980) *Dynamics of Synergetic Systems*, Springer, Berlin-Heidelberg-New York.

Heisenberg, W. (1959) *Wandlungen in dem Grundlagen der Naturwissenschaft*, S. Hirzel Verlag, Stuttgart (9th ed.). (English trans. in W. HEISENBERG, *Collected Works*, Springer, Berlin—Heidelberg—New York).

Hempel, C. G. (1942) *The Function of Laws in History*, "Journal of Philosophy", XXXIX.

Hess, B.—Goldbeter, A.—Lefever, R. (1978) *Temporal, Spatial, and Function Order in Regulated Biochemical and Cellular Systems*, "Advances in Chemical Physics", XXXVIII, pp. 363–413.

Hess, B.—Markus, M. in E. Frehland (ed.) (1984) *Synergetics—From Microscopic to Macroscopic Order*, Springer, Berlin-Heidelberg-New York.

Heyer, P. (1982) *Nature, Human Nature and Society: Marx, Darwin, Biology and Human Sciences*, Greenwood Press, London.

Ho, M. W.—Saunders, P. T. (1979) *Beyond neo-Darwinism. An epigenetic approach to evolution*, "Journal of Theoretical Biology", 78, pp. 573–91.

Hobsbawm, E. J. (1962) *The age of Revolution. Europe 1789–1848*, London.

Hobsbawm, E. (1980) *The Revival of Narrative: Some Comments*, "Past and Present", n.86.

Horkheimer, M. (1930) *Aufänge der bürgerlichen Geschichtsphilosophie*, Kohlhammer, Stuttgart.

Horkheimer, M.—Adorno, T. W. (1947) *Dialektik der Aufklärung, Philosophische Fragmente*, Querido Verlag, Amsterdam. (English trans. *Dialetic of Enlightenment*, Continuum, New York, 1975).

Horsthemke, W. (1980) *Non Equilibrium transitions Induced by External White and Coloured Noise*, in H. HAKEN (ed.) *Dynamics of Synergetic Systems*, Springer, Berlin-Heidelberger-New York.

Hull, D. L. (1974) *Philosophy of Biological Science*, Prentice-Hall, Englewood Cliffs (N.J.).

Husserl, E. (1954) *Die Krisis der europäischen Wissenschaftens und die transzendentale Phänomenologie*, Martinus Nijhoff, Haag (Eng. trans. *Crisis of European Sciences and Transcendental Phenomenology*, Northwestern University Press, Evanston, 1970).

Jacob, F. (1970) *La Logique du vivant. Une historie de l'hérédité*, Gallimard, Paris (English trans., *The logic of Life: A History of Heredity*, Pantheon Books, New York, 1982).

Jacob, F. (1974) *Evolution et realisme*, "Publications de l'Université de Lausanne", fasc. XLIV, pp.21–34.

Jacob, F. (1977) *Evolution and tinkering*, in "Science", 196, pp.1161–66.

Jacob, F. (1983) *Molecular tinkering in evolution*, in D. S. BENDALL (ed.), *Evolution from molecules to men*, Cambridge University Press, Cambridge.

James, W. (1890) *Principles of Psychology*, Henry Holt and Company.

Jantsch, E. (1975) *Design for Evolution*, Braziller, New York.

Jantsch, E. (1980) *The Self-Organizing Universe*, Pergamon Press, Oxford.

Jantsch, E. (1981) *The Evolutionary Vision*, Westview Press, Boulder (Colo).

Jantsch, E.—Waddington, C. H. (eds.) (1976) *Evolution and Consciousness*, Addison—Wesley, Reading (Mass.).

Jung, C. G. (1954) *Theoretische Überlegungen zum wesen des Psychischen* in *Von den Wurzeln des Bewüsstseins*, Roscher, Zurich (English trans. *On the Nature of the Psyche*, Princeton University Press, Princeton (N.J.) 1969).

Jung, C. G.—Pauli, W. (1952) *Naturerklärung und Psyche*, Zürich (English trans. *The Interpretation of Nature and Psyche*, Routledge & Kegan Paul, London, 1955).

Jung, C. G. et al. (1964) *Man and his symbols*, Aldus Books, London.

Katchalsky, A.—Curran, P. F. (1965) *Nonequilibrium Thermodynamics in Biophysics*, Cambridge (Mass.).

Kittsteiner, H. D. (1985) *Il concetto di relatività nelle scienze storiche*, Paper given at Congress on *Einstein e il suo tempo*, Venezia, December 1985.

Koestler, A. (1967) *The Ghost in the Machine*, Macmillan, London.

Krader, L. (1968) *Critique dialectique de la nature humaine*, "L'homme et la societé", n.10.

Kuhn, T. S. (1962) *The Structure of Scientific Revolutions*, University of Chicago Press, Chicago.

Laszlo, E. (1973) *Introduction to Systems Philosophy*, Harper Torchbooks, New York.

Laszlo, E. (1986) *Evoluzione*, Feltrinelli, Milano. (English edition revised and enlarged, *Evolution: The Grand Synthesis*, New Science Library, Shambhala, Boston-London, 1987).

Lakatos, I., Musgrave, A. (eds.) (1970) *Criticism and the Growth of Knowledge*, Cambridge University Press, Cambridge.

Lefever, R.—Deneubourg, J. L. (1975) *On the changes in Conductance and Stability properties of electrically excitable membranes during voltage-clamp experiments*, "Advances in Chemical Physics", volume XXIX, pp. 349–74.

Leff, G. (1969) *History and Social Theory*, Garden City (N.Y.).

Le Moigne, J. L. (1982) *Systematique et épistémiologie*, in J. LESOURNE (ed.) *La Notion de Système dans les Sciences Contemporaines*, Librairie de l'Université, Aix en Provence, tome II,

Lesnoff, M. (1974) *The Structure of Social Science*, London.

Lewontin, R. (1977) *Adattamento*, in *Enciclopedia Einaudi*, vol. I, Torino, pp. 198–214.

Lewontin, R. (1978) *Adaptation*, "Scientific American", 239, (9), pp.156–69.

Lewontin, R. (1979) *Sociology as an adaptationist program*, "Behavioral Science", 24, pp. 5–14.

Lewontin, R. (1983) *Gene, organism and environment*, in D. S. BENDALL, (ed.) *Evolution from molecules to men*, Cambridge University Press, Cambridge.

Lewontin, R.—Levins, R. (1978) *Evoluzione*, in *Enciclopedia Einaudi*, Vol. 5, Torino, pp. 995–1051.

Livingstone, P. (ed.) (1984) *Disorder and Order*, Anma Libri, Stanford.

Lombardo Radice, L. (1978) *Prefazione* a F. Engels, *Dialettica della natura*, Editori Riuniti, Roma.

Lukács, G. (1969) *Die ontologischen Grundlagen des menschlichen Denkens und Handelns*, "Ad. lectores", No. 8.

Martin, R. (1977) *Historical Explanation, Re-enactment and pratical Inference*, Ithaca and London.

Marx, K. (1844) *Oekonomisch-philosophische Manuscripte aus dem Jahre 1844*, in K. MARX—F. ENGELS, *Gesamtansgabe*, Abt.1, Bd.3, Berlin, 1932. (English trans. *Economics and Philosophic Manuscripts of 1844*, in K. MARX—F. ENGELS, *Collected Works*, volume 3, International Publishers, New York, 1975).

Maturana, H.—Varela, F. (1980) *Autopoiesis and Cognition. The Realisation of the Living*, Reidel, Dordrecht.

Maturana, H.—Varela, F. (1985) *The Tree of Knowledge*, New Science Library, Boston.

Mayr, E. (1982) *The Growth of biological Thought*, Harvard University Press, Cambridge (Mass.).

Monod, J. (1970) *Le hasard et la nécessité. Essai sur la philosophie naturelle de la biologie moderne*, Seuil, Paris. (English trans. *Chance and Necessity*, Random House, New York, 1972).

Morin, E. (1977) *La Méthode. I. La Nature de la Nature*, Seuil, Paris.

Morin, E. (1980) *La Méthode. II. La Vie de la Vie*, Seuil, Paris.

Morin, E. (1984) *Sur la définition de la complexité*, Communication au colloque *Science et Pratique de la complexité*, The United Nation University, Montpellier, mai 1984.

Morowitz, H. (1968) *Energy Flow in Biology*, Academic Press, New York.

Muller, G. H. (1983) *Darwin, Marx, Aveling-Briefe und Spekulationen. Eine bibliographische Betrachtung*, "Dialektik", 6, pp.149–159.

Nagel, E. (1952) *Some Issues in the Logic of Historical Analysis*, "Scientific Monthly", LXXIV.

Nardone, G. (1971) *Il pensiero di Gramsci*, Laterza, Bari.

Needham, J. (1968) *Order and Life*, New edit., MIT Press, Cambridge (Mass.).

Nicolis, G—Prigogine, I. (1977) *Self-Organization in Non-Equilibrium Systems*, Wiley Interscience, New York.

Nietzsche, F. (1874) *Unzeitgemasse Betrachtungen, Zweits Stuck: Vom Nutzen und Nachteil der Histoire fur das Leben*, E. W. Fritzsch, Leipziy (English trans. *Thoughts out of Season*, part two, *The Use and Abuse of History*, T. N. Foulis, Edinburgh and London, 1909).

O'Malley, J. J. (1966) *History and, Man's "Nature" in Marx*, "The Review of Politics", n.4.

Onsager, L. (1931) *Reciprocal Relations in Irreversible Processes*, "Physiological Review", 38.

Parsons, T. (1966) *Societes* in *Evolutionary and Comparative Perspective*, Prentice-Hall, Englewood Cliffs (N.Y.).

Patte, H. (ed.) (1973) *Hierarchy Theory: the Challenge of Complex Systems*, Braziller, New York.

Pauli, W. (1961) *Aufsätze und Vortäge über Physik und Erkenntnistheorie*, Verlag Vieweg, Braunschweig.

Pauli, W. (1979) *Wissenschaftlicher Briefwechsel mit Bohr, Einstein, Heisenberg, U.A.*, Band I: 1919–1929; *Scientific Correspondence with Bohr, Einstein, Heisenberg, A.O.*, Volume I: 1919–1929, Springer-Verlag, New York-Heidelberg-Berlin.

Piaget, J. (1967) *Biologie et connaissance*, Gallimard, Paris (English trans. *Biology and Knowledge*, Edinburg University Press, 1971).

Piaget, J. (1970) *L'épistémiologie génétique*, PUF, Paris (English trans. *Principles of Genetic Epistemiology*, Routledge.

Piaget, J. (1974) *Adaptation vitale et psychologie de l'intelligence. Sélection organique et phénocopie*, Hermann, Paris (English trans. *Adaptation and Intelligence: Organic Selection and Phenocopy*, University of Chicago Press.

Piaget, J. (1975) *L'équilibration des structures cognitives. Problème central du developpement*, PUF, Paris (English trans. *The equilibration of congitive structures: The central problem of intellectual development*, University of Chicago Press.

Piaget, J. (1976) *Le comportement, moteur de l'evolution*, Gallimard, Paris (English trans. *Behaviour and Evolution*, Routledge.

Piaget, J.—Garcia, R. (1983) *Psychogenèse et histoire des sciences*, Flammarion, Paris

Pomian, K. (1977) *Ciclo*, in *Enciclopedia Einaudi*, Torino, vol. 2.

Prestipino, G. (1973) *Natura e società. Per una nuova lettura di Engels*, Editori Riuniti, Roma.

Prigogine, I. (1947) *Etude Thermodynamique des Phenomènes Irreversibles*, Desoer, Liège.

Prigogine, I.—Lefever, R. (1975) *Stability and Self-Organisation in Open Systems*, "Advances in Chemical Physics", volume XXIX, pp.1–28.

Prigogine, I.—Stengers, I. (1979) *La Nouvelle Alliance. Métamorphose de la science*, Gallimard, Paris. (English ed.: *Order out of Chaos. Man's New Dialogue with Nature*, Flamingo, Fontana Paperbacks, London, 1985; cf. also Italian ed.: *La Nuova alleanza. Metamorfosi della scienza*, Einaudi, Torino, 1981).

Prigogine, I. (1980) *From being to becoming*, W. H. Freeman and Company, New York.

Rasmusen, L. B. (1975) *Two Essays on the Scientific Study of History*, Bern and Frankfurt.

Richter, P. H. in H. Hacken (ed.) (1979) *Pattern Formation by Dynamic Systems and Pattern Recognition* Springer, Berlin, Heidelberg, New York, pp.155–65.

Rickert, H. (1896–1902) *Die Grenzer der naturwissenschaftlichen Begriffsbildung*, (4th. ed. Tubingen, 1921). (English trans., *Limits of Concept Formation in Natural Sciences*, Cambridge University Press, Cambridge).

Riedel, M. (1969) *Studien zu Hegels Rechtsphilosophie*, Suhrkamp, Frankfurt (English trans. *Between Tradition and Revolution. The Hegelian Transformation of Political Philosophy*, Cambridge University Press, Cambridge 1984).

Romano, R. (ed.) (1981) *Le frontiere del tempo*, Il Saggiatore, Milano.

Rossi, Paolo (1981) *Fatti scientifici e stili di pensiero: appunti attorno ad una rivoluzione immaginaria*, "Rivista di filosofia", 21, pp.403–28.

Rossi, Pietro (ed.) (1970) *Lo storicismo contemporaneo*, Loescher, Torino.

Rusconi, G. E. (1970₂) *La teoria critica della società*, Il Mulino, Bologna.

Russel, B. (1903) *The Principles of Mathematics*, Cambridge University Press, Cambridge.

Sabetti, A. (1958) *Il rapporto uomo-natura nel pensiero di Gramsci e la fondazione della scienza*, in Varions Authors, *Studi gramsciani*, Editori Riuniti, Roma.

Schrödinger, E. (1956) *Expanding Universes*, Cambridge University Press, Cambridge.

Schuster, P. in E. FREHLAND (ed.) (1984) *Synergetics From Microscopic to Macroscopic Order*, Springer, Berlin-Heidelberg-New York.

Serres, M. (1977) *La naissance de la physique dans le texte de Lucrèce*, Les Editios de Minuit, Paris.

Serwer, D. (1977) *Unmechanischer Zwang: Pauli, Heisenberg and the Rejection of the Mechanical Atom, 1923–1925*, "Historical Studies in the Physical Sciences", vol. III, pp. 189–256.

Stanley, S. M. (1975) *A Theory of Evolution Above the Species Level*, in "Proceedings of the National Academy of Sciences USA", 72.

Stanley, S. M. (1981) *The New Evolutionary Timetable*, Basic Books, New York.

Stedman Jones, G. (1972) *Down with Nature*, "7 Days", January.

Stone, L. (1979) *The Revival of Narrative: Reflections on a New Old History*, "Past and Present", n.85.

Thom, R. (1972) *Stabilité structurelle et Morphogénèse*, InterEditions, Paris. (Engl. trans. *Structural Stability and Morphogenesis*, Benjamin-Cunsmings, Menlo Park (CA).

Thom, R. (1980) *Parabole e catastrofi*, Il Saggiatore, Milano.

Thompson, E. P. (1978) *The poverty of theory and Other Essays*, Merlin Press, London.

Timpanaro, S. (1970) *Sul materialismo*, Nistri-Lischi, Pisa.

Topolski, J. (1983) *The Scientific Character of Historiography and its Limits*, in Pietro ROSSI (ed.) *La teoria della storiografia oggi*, Il Saggiatore, Milano, 1983.

Toraldo di Francia, G. (1976) *L'indagine del mondo fisico*, Einaudi, Torino.

Trotsky, L. (1930) *My Life. The rise and fall of a dictator*, Thornton Butterworh, London.

Vaiani, C. (1971) *L'Abbild-theorie in Gyorgy Lukács*, L'Aquila.

Varela, F. (1979) *Principles of biological Autonomy*, North Holland, New York.

Varela, F. (1984a) *Living Ways of Sense-Making: A Middle Path for Neurosciences* in P. LIVINGSTONE, (ed.) *Disorder and Order*, Anma Libri, Stanford.

Varela, F. (1984b) *The Creative Circle: Sketches on the Natural History of Circularity*, in P. WATZLAWICK (ed.), *The Invented Reality*, Norton, New York.

Varela, F.—Goguen, J. (1978) *The Arithmetics of closure*, in R. TRAPPL (ed.) *Progress in Cybernetics and Systems Research*, vol. III, Hemisphere Publ. Co., Washington.

Veyne, P. (1971) *Comment on écrit l'histoire. Essai d'epistemiologie*, Editions du Seuil, Paris. (English trans. *Writing History: Essay on Epistemiology*, Manchester University Press).

Vidoni, F. (1985) *Natura e storia. Marx ed Engels interpreti del darwinismo*, Edizioni Dedalo, Bari.

Waddington, C. H. (1975) *The Evolution of an Evolutionist*, Edinburgh University Press, Edinburgh.

Watzlawick, P. (ed.) (1981) *Die erfundene Wirklichkeit*, Piper, München (English trans. *The Invented Reality*, Norton, New York 1984.

Weber, M. (1904) *Die "Objektivität" sozialwissenschaftlicher und sozialpolitischer Erkenntnis*, in *Gesammelte Aufsätze zur Wissenschaftslehre*, Mohr, Tübingen, 1922. (English trans. *The Methodology of Social Science*, The Free Press, Glencol (Ill.) 1949).

Weber, M. (1903–6) *Roscher und Knies und die logischen Probleme der historischen Nationalökonomie*, in *Ges. Aufs. zur Wissenschaftslehre*.

Weber, M. (1917) *Der Sinn der "Wertfreiheit" der soziologischen und ökonomischen Wissenschaftslehre*, in *Ges. Aufs. zur Wissenshaftslehre*.

Weiss, P. (1969) *The Living System; Determinism Stratified*, in A. KOESTLER-J. R. SMYTHIES (eds.), *Beyond Reductionism*, Hutchinson, London.

Weiss, P. et al. (1971) *Hierarchically Organized Systems in Theory and Practice*, Hafner, New York.

Weiss, P. (1974) *L'archipel scientifique*, Maloine, Paris.

White, H. (1973) *Metahistory, The Historical Imagination in Ninetheenth-Century Europe*. The Johns Hopkins University Press, Baltimore-London.

White, H. (1978) *Topics of discourse, Essays in Cultural Criticism*, The Johns Hopkins University Press, Baltimore-London.

Whitehead, A. N. (1954) *Adventures of Ideas*, Macmillan, New York.

Windelband, W. (1894) *Geschichte und Naturwissenschaft*, in *Praludien*, vol. II (5th. ed., Tubingen, 1915).

von Wright, G. H. (1971) *Explanation and understanding*, Cornell University Press, Ithaca (N.Y.).

Young, J. Z. (1971) *An Introduction to the Study of Man*, Oxford University Press, Oxford.

Yovitz, M. C.—Jacobi, G. T.—Goldstein, G. D. (1962) *Self-organizing systems*, Spartan Books, Washington.

Zeeman, Christopher (1977) *Catastrophe Theory*, Benjamin, Reading (Mass).

Zeleny, M. (ed.) (1980) *Autopoiesis, Dissipative Structures and Spontaneous Social Orders*, Westview Press, Boulder (Col.).

Zeleny, M. (ed.) (1981) *Autopoiesis. A Theory of Living Organization*, North Holland, New York.

BIBLIOGRAPHY TO PART TWO

Arbizzani, L. (1984) *Lotte e organizzazioni contadine nell'ultimo ventennio dell'Ottocento nelle province orientali dell'Emilia Romagna*, in Various Authors, *Rivolte e movimenti contadini nella Valle Padana di fine Ottocento*, "Annali dell'Istituto Cervi", Volume VI, pp.203 et seg.

Armengaud, A. (1973) *Population in Europe 1700–1914*, in C. M. CIPOLLA (ed.), *The Fontana Economic History of Europe*, volume 3, *The industrial revolution*, (new edition: Barnes and Nolte New York, 1976).

Artigiani, R. (1987) *Cultural Evolution*, "World Furtures", volume 23, pp. 93–121.

Aspinall, A. (1949a) *The Early English Trade Unions*, Batchworth, London.

Aspinall, A. (1949b) *Politics and the Press*, Batchworth, London.

Barbadoro, I. (1973) *Storia del sindacalismo italiano dalla nascita al fascismo*, volume I, *La Federterra*, La Nuova Italia, Firenze.

Berend, I. T.—Ránki, G. (1974a) *Economic Development in East-Central Europe in the 19th and 20th Centuries*, Columbia University Press, New York-London.

Berend, I. T.—Ránki, G. (1974b) *Hungary. A Century of Economic Development*, Harper and Row, New York.

Cafagna, L. (1961) *La formazione di una base industriale fra il 1896 ed il 1914*, "Studi storici", Nos. 3–4, pp. 690–724.

Capitini Maccabruni, N. (1964) *Gli scioperi delle trecciaiole toscane del 1896–97 e l'azione delle Camere del Lavoro di Firenze*, "Movimento operaio e socialista", No.2, pp.121 et seq.

Carocci, G. (1956) *Agostino Depretis e la politica interna italiana dal 1876 al 1887*, Einaudi, Torino.

Cole, G. D. H. (1948) *A Short History of the British Working-Class Movement 1789–1947*, George Allen & Unwin, London.

Conze, W. (1957) *Die Strukturgeschichte des technisch-industriellen Zeitalters*, Köln.

Conze, W.—Engelhardt V. (Hrsg) (1979) *Arbeiter im Industrialisierungsprozess Herkunft, Lage und Verhalten*, Stuttgart.

Della Peruta, F. (1984) *Il movimento contadino nell'alto Milanese (1885–1899)*, in Various Authors, *Rivolte e movimenti contadini nella Valle Padana di fine Ottocento*, "Annali dell'Istituto Cervi", Volume VI, pp.41 et seq.

Darvall, F. O. (1934) *Popular disturbances and public order in Regency England*, London.

Fischer, W. (1972) *Wirtschaft und Gesallschaft im Zeitalter der Industrialisie-rung*, Göttingen.

Forti, C. (1975) *Le leghe contadine mantovane dal 1898 allo sciopero del 1904*, in Various Authors, *Braccianti e contadini nella Valle Padana*, Editori Riuniti, Roma, pp.381–456.

Fossati, A. (1951) *Lavoro e produzione in Italia dalla metà del secolo XVIII alla seconda guerra mondiale*, Giappichelli, Torino.

Fraser, W. H. (1971) *Robert Owen and the Workers*, in J. BUTT (ed.), *Robert Owen, A Symposium*, David & Charles, Newton Abbot.

Genzini Romani, V. (1975) *Il movimento contadino nel Cremonese all'inizio del '900*, in Various Authors, *Braccianti e contadini nella Valle Padana*, Editori Riuniti, Roma, pp.87–134.

Gerschenkron, A. (1962) *Economic Backwardness in Historical Perspective*, Harvard University Press, Cambridge (Mass).

Giacobbi, S. (1975) *Agricoltura e contadini nel Cremonese dall'Unità alla fine del secolo*, in Various Authors, *Braccianti e contadini nella Valle Padana*, Editori Riuniti, Roma, pp.1–86.

Gould, J. D. (1979) *European Inter-Continental Emigration 1815–1914: Pattern and Causes*, "Journal of European Economic History", No.3.

Hammond, J. L.—B. (1919) *The Skilled Labourer 1760–1832*, Longman, London.

Haupt, G. (1978) *Perché la storia del movimento operario?*, in ID., L'internazionale socialista dalla Comune a Lenin, Einaudi, Torino.

Hobsbawm, E. J. (1952) *The Machine Breakers*, "Past and Present", I, pp.52–70.

Hobsbawm, E. J. (1964) *Labouring Men. Studies in the History of Labour*, Weidenfeld and Nicolson, London.

Hobsbawm, E. J. (1971) *From Social History to the History of Society*, "Daedalus".

Hobsbawm, E. J. (1974) *Labour History and Ideology*, "Journal of Social History", Summer, pp.371 et seq.

Hobsbawm, E. J. (1984) *Worlds of Labour*, Weidenfeld and Nicolson, London.

Inchiesta Agraria (1881) *Atti della Giunta per l'Inchiesta agraria e sulle condizioni della classe agricola*, volume II, fasc. I, Roma.

Kocka, J. (1969) *Unternehmensverwaltung und Angestelltenschaft am Beispiel Siemens 1847–1914*, Stuttgart.

Landes, D. S. (1969) *The Unbound Promethus. Tecnological change and industrial development in Western Europe from 1750 to the present*, Cambridge University Press, Cambridge (2nd ed. 1972).

Laszlo, E. (1986) *Evoluzione*, Feltrinelli, Milano. (English edition, revised and enlarged, *Evolution: the Grand Synthesis*, New Science Libary, Shambhala, Boston-London, 1987).

Lenin, V. I. (1917) *Agranaya programma sotsial-democratii v pervoi russkoi: revolutsii 1905–1097 gg.*, Zhizn i Znaniye. (English trans: *The Agrarian Program of the Social-Democrats in the First Russian Revolution of 1905–1907*, in *Collected Works*, volume 13, Foreign Languages Publishing House, Moscow, 1962).

Léon, P. (1978) *Histoire économique et sociale du monde*, tome 4, *La domination du capitalisme. 1840–1914*, Librairie Arman Colin, Paris.

Lequin, Y. (1977) *Les ouvriers de la région lyonnaise (1848–1914)*, Presses Universitaires de Lyon, Lyon.

Lyashchenko, P. I. (1949) *History of the National Economy of Russia to the 1917 Revolution*, The Macmillan Company, New York.

Luzzatto, G. (1968) *L'economia italiana dal 1861 al 1894*, Einaudi, Torino.

Malefakis, E. E. (1970) *Agrarian Reform and Peasant Revolution in Spain*, Yale University Press, New Haven and London.

Masulli, I. (1980) *Crisi e trasformazione: strutture economiche, rapporti sociali e lotte politiche nel Bolognese (1880–1914)*, Istituto per la storia di Bologna, Bologna.

Medici, G.—Orlando, G. (1952) *Agricoltura e disoccupazione. I braccianti della bassa pianura padana*, Bologna.

Merli, S. (1972) *Proletariato di fabbrica e capitalismo industriale. Il caso italiano 1880–1900*, La Nuova Italia, Firenze (ristampa 1976).

Merriman, J. (ed.) (1979) *Consciousness and Class Experience in nineteenth Century Europe*, Holmes and Meier Publishers, New York.

Ministero di Agricoltura, Industria e Commercio (1894–1902) *Statistica degli scioperi avvenuti nell'Industria e nell'Agricoltura durante l'anno 1892 . . . 1893 . . . 1897 . . . 1898 . . . 1899 . . . 1900*, Roma.

Musson, A. E. (1972) *British Trade Unions 1800–1875*, Macmillan, London.

Oliver, W. R. (1958) *Robert Owen and the English Working Class Movements*, "History today".

Perrot, M. (1974) *Les ouvriers en grève (France, 1871–1890)*, Mouton, Paris.

Perrot, M. (1978) *Les ouvriers et les machines en France dans la première moitié du XIX siècle*, "Recherches", Nos. 32–33.

Prigogine, I.—Stengers, I. (1979) *La Nouvelle Alliance. Métamporphose de la science*, Gallimard Paris. (English edition: *Order out of chaos. Man's new Dialogue with Nature*, Flamingo, Fontana Paperbacks, London, 1986).

Ramella, F. (1975) *Il problema della formazione della classe operaia in Italia. Il caso di un distretto industriale dell'Ottocento*, "Classe", No. 10; pp. 107–125.

Reinhard, M. R.—Armengaud, A.—Dupaquier, J. (1968) *Histoire générale de la population mondiale*, Editions Montchrestien, Paris.

Roveri, A. (1972) *Dal sindacalismo rivoluzionario al fascismo. Capitalismo agrario e socialismo nel Ferrarese (1870–1920)*, La Nuova Italia, Firenze.

Rude', G. (1981) *The Crowd in history, 1730–1848*, Lawrence and Wishart, London.

Salvati, M. (1980) *Cultura operaia e disciplina industriale: ipotesi per un confronto tra correnti storiografiche*, "Movimento operaio e socialista", No. 1, pp.5–17.

Samuel, R. (1977) *The Workshop of the World: Steam Power and Hand Technology in mid-Victorian Britain*, "History Workshop", No. 3.

Schomerus, H. (1977) *Die Arbeiter der Maschinenfabrik Esslingen*, Stuttgart.

Schumpeter, J. A. (1939) *Business Cycles. A theoretical, Historical and Statistical Analysis of the Capitalist Processes*, McGraw-Hill Book Company, New York.

Sereni, E. (1966) *Capitalismo e mercato nazionale in Italia*, Editori Riuniti, Roma.

Sori, E. (1981) *Mercato del lavoro ed emigrazione*, in N. TRANFAGLIA (ed.) *Il mondo contemporaneo*, vol. II, *Storia d'Europa*, t. IV, La Nuova Italia, Firenze.

Stedman Jones, G. (1974) *Working-class culture and Working-class politics in London 1870–1900: notes on the remaking of working class*, "Journal of Social History", No. 4.

Tenfelde, K. (1977) *Sozialgeschichte der Bergarbeiterschaft an der Ruhr im 19, Jahrhundert*, Bon Bad Godesberg.

Thompson, E. P. (1963) *The Making of the English Working Class*, (reprint: Penguin Books, Harmondsworth, 1979).

Thompson, E. P. (1967) *Time, Work-discipline and industrial capitalism*, "Past and Present", No. 38.

Tilly, Ch.—L.—R. (1975) *The Rebellious Century, 1830–1930*, Dent & S., London.

Trempe', R. (1971) *Les mineurs de Carmaux*, Edition Ouvrières, Paris.

Turner, H. A. (1962) *Trade Union Growth, Structure and Policy. A Comparative Study of the Cotton Unions*, Allen & Unwin, London.

Ufficio del Lavoro della Societa' Umanitaria (1904) *La disoccupazione nel basso emiliano. Inchiesta diretta nelle province di Ferrara, Bologna e Ravenna*, Milano.

Valenti, G. (1912) *L'Italia agricola dal 1861 al 1911*, in Various Authors, *Cinquant'anni della storia italiana*, Accademia dei Lincei, Roma.

Various Authors (1974) *Zwischen Sozialgeschichte und Legitimationswissenschaft*, "Arbeiterbewegung, Theorie und Geschichte. Jahrbuch 2".

Various Authors (1977) *Making and Change of Plebeian and Proletarian Consciousness*, "7th Round table in Social History", University of Constance.

Various Authors (1978) *Storia sociale e storia del movimento operaio*, "Annali della Fondazione Basso", volume IV, 1978–80.

Villari, L. (1965) *Per una ricerca sulla storia del protezionismo in Italia*, "Studi storici", Nos. 3–4.

Vitali, O. (1970) *Aspetti dello sviluppo economico italiano alla luce della ricostruzione della popolazione attiva*, Roma.

Webb, S.—B. (1911) *The History of Trade Unionism*, Longmans—Green & Co., London.

Zangheri, R. (1960) *Lotte agrarie in Italia. La Federazione nazionale dei lavoratori della terra 1901–1926*, Feltrinelli, Milano.